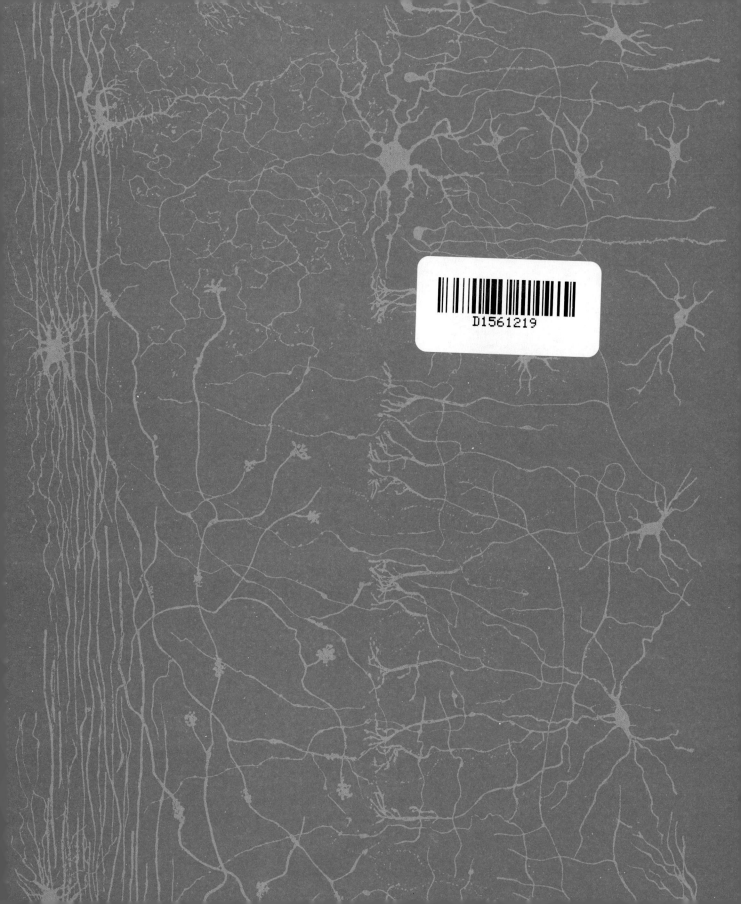

THE CELL

THE CELL
A VISUAL TOUR OF THE BUILDING BLOCK OF LIFE

JACK CHALLONER

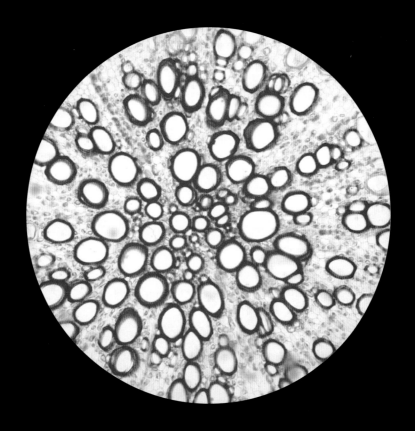

CONSULTANT EDITOR
DR PHIL DASH

THE UNIVERSITY OF CHICAGO PRESS
Chicago and London

Jack Challoner is the author of more than thirty books on science and technology. He also works as an independent science consultant for print, radio, and TV.

The University of Chicago Press, Chicago 60637
The University of Chicago Press, Ltd., London
© 2015 by The Ivy Press Limited
All rights reserved. Published 2015.
Printed in China

24 23 22 21 20 19 18 17 16 15   1 2 3 4 5

Text © Jack Challoner 2015
Design and layout © The Ivy Press Limited 2015

ISBN-13: 978-0-226-22418-3 (cloth)
ISBN-13: 978-0-226-22421-3 (e-book)
DOI: 10.7208/chicago/978-0226224213.001.0001

Library of Congress Cataloging-in-Publication Data
Challoner, Jack, author.
The cell : the origin of life / Jack Challoner.
       pages cm
Includes index.
ISBN 978-0-226-22418-3 (cloth : alk. paper) — ISBN 978-0-226-22421-3 (ebook) 1. Cells. 2. Cytology. I. Title.
  QH582.4.C425 2015
  571.6—dc23   2015008716

This book was conceived, designed, and produced by
Ivy Press
210 High Street, Lewes, East Sussex, BN7 2NS. United Kingdom
www.ivypress.co.uk

Publisher Susan Kelly
Creative Director Michael Whitehead
Editorial Director Tom Kitch
Art Director James Lawrence
Commissioning Editor Jacqui Sayers
Project Editor Stephanie Evans
Designer Andrew Milne
Illustrator Vivien Martineau
Picture Researcher Caroline Hensman

Color origination by Ivy Press Reprographics

∞ This paper meets the requirements of ANSI/NISO Z39.48-1992 (Permanence of Paper).

Front cover image Spike Walker, Wellcome Images.
Back cover image Lebendkulturen.de/ Shutterstock, Inc.

Contents

6 | INTRODUCTION
Why the cell is Earth's greatest success story, and the basis of all life.

10 | CHAPTER 1
A Brief History of the Cell
In the 350 years since cells were discovered remarkable progress has been made in our understanding of them.

58 | CHAPTER 3
Cells Beget Cells
The process of cell division accounts for growth and reproduction, as well as the evolution of new species.

28 | CHAPTER 2
Inside Living Cells
All types of cell share certain characteristics, including the molecular machinery that makes them work.

90 | CHAPTER 4
Cellular Singletons
The overwhelming majority of cells on Earth are individual living things—single-celled organisms.

142 | CHAPTER 6
Life, Death, and Immortality
Cells have evolved extraordinary ways to attack other cells and to protect themselves.

118 | CHAPTER 5
Coming Together—Multicellular Life
Cells cooperate within complex organisms and perform essential specialized tasks.

164 | CHAPTER 7
Taking in the Cytes
The human body manufactures around 200 different cell types, displaying astonishing diversity and specialization.

188 | GLOSSARY
INDEX
ACKNOWLEDGMENTS

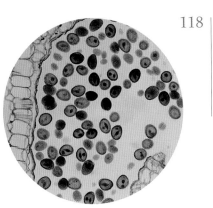

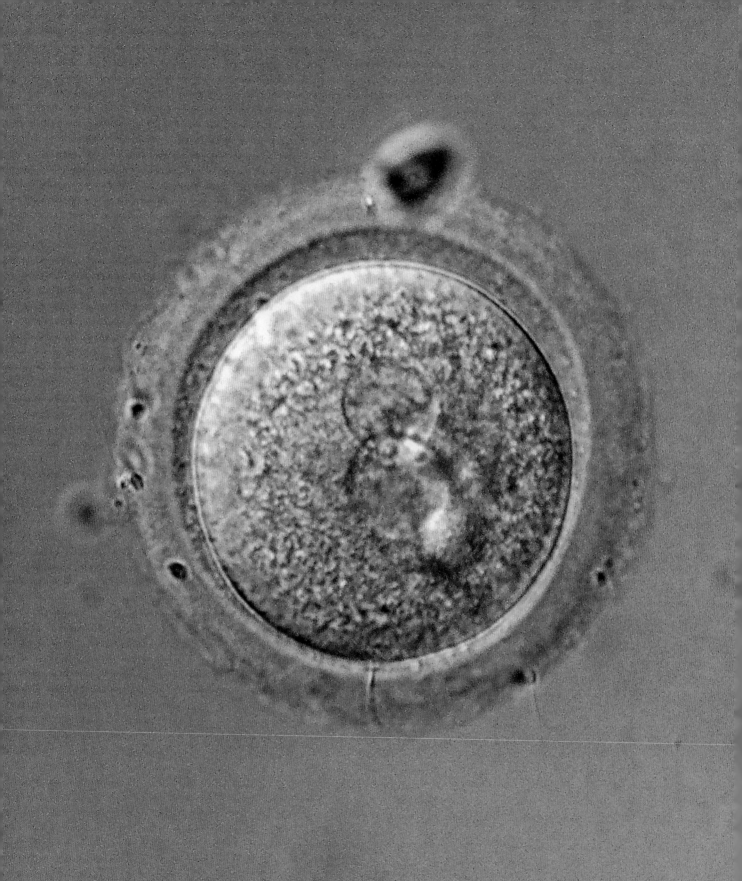

Introduction

Every person on this planet began life as one cell, about the same size as this period. Each one of us remained like this for about 24 hours before dividing in two—the first step toward creating the complex, multicellular organism that humans are today. It is an incredible and fascinating thought, that a human could have been contained in a single cell, and that, perhaps more remarkable still, that one basic unit of life knew what to do to next.

To understand just how important cells are, consider the following. The total number of living things currently inhabiting our planet is unimaginably large (there are an estimated 8.7 million unique species, most of them numbering millions, billions, or trillions of individuals), and every last one of them, without exception, is made of one or more cells. Next, consider the incredible variety of processes and materials that occur in the natural world. The glow of a firefly, a plant bending toward the light, cancer, a 100-meter sprint, wood, mucus, elephant dung, a blue whale's skeleton, body odor, the memory of the smell of ratatouille, the call of a howler monkey, houseplants, a hawk's beak, a snake's venom … all of these are the result of activity in cells.

What is this life?

The difference between a living and a nonliving thing has always been difficult to define. Biologists generally agree that for something to be considered alive it must satisfy a set of criteria, including the use of energy to build complex molecules and organize its internal systems, and the ability to respond to its surroundings and to reproduce. Rhododendrons and ants satisfy all of these criteria—but only because they are made of cells, life's building blocks. Cells are life, and to understand their behavior, their structure and their remarkable microscopic and submicroscopic machinery is to understand life itself. Chapter one outlines the history of that understanding (so far), and examines some of the tools and techniques that have nurtured it.

A cell is just a mixture of molecules, a cocktail of chemicals, inside a little bag. Despite the unambitious simplicity of that description,

Left *Two become one; light micrograph (a microscope photograph) of a single, fertilized human egg cell (ovum) soon before the first cell division.*

Below *Anything at all in the living world, from a material to a process, happens because of activity in cells.*

and the tiny size of a typical cell, truly intricate wonders lie within. This is the subject of chapter two, the inside view of cells, in which the main features are identified and the main types of cell are considered.

The cell is a bewilderingly busy molecular metropolis: some molecules are making copies of themselves; some are manufacturing other molecules; some are quickly reading information coded along the length of others; some are grabbing hold of others and carrying them to wherever they need to be next; some are self-assembling as a kind of molecular scaffolding or as tracks along which other molecules can be carried. All of this dizzying activity, and infinitely more, is taking place in every living cell on the planet every moment of every day. Two of the most important outcomes of this molecular dance—growth and reproduction, via the proliferation of cells—form the subject of chapter three.

How big are cells?

Although most individual cells are far too small to see without serious magnification, there are some that are big enough. Bird eggs, for example, are single cells. The largest bird egg of a living species is laid by the ostrich, and ostrich eggs are, in fact, the largest of all cells. (The shell, incidentally, is not part of the cell but is manufactured by it.) There are millions of species that remain as single cells all their lives— the ostrich is not among them, of course. Among the largest of them is *Valonia ventricosa*, also known as bubble algae, which can grow to 2 inches (5 centimeters) in diameter. The variety and importance of these cellular singletons is revealed in chapter four.

The living things with which we are familiar—the ones we can actually see—are made of thousands, millions, billions, or trillions of cells. Nearly all multicellular organisms are either plants, fungi, or animals. In most cases the cells that make up these organisms all come from repeated cell divisions that

HOW STRANGE THE CHANGE FROM MICRO TO MACRO

Most cells are too small to be seen with the naked eye, but they are easily observed with light microscopes; they are microscopic. The smallest cells, tiny mycoplasma that live inside other cells, are under one thousandth of a millimeter (1 micron) in diameter, while the largest, such as bird eggs or nerve cells, are inches across. Most types of cell are typically between 5 and 10 microns across.

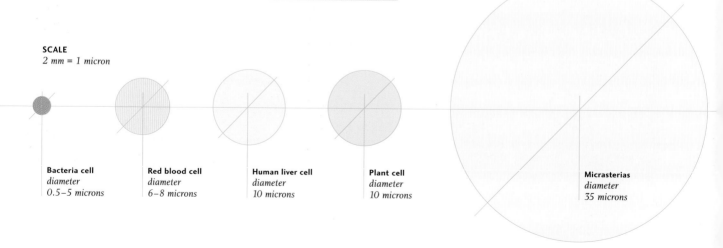

SCALE
2 mm = 1 micron

Bacteria cell
diameter
0.5–5 microns

Red blood cell
diameter
6–8 microns

Human liver cell
diameter
10 microns

Plant cell
diameter
10 microns

Micrasterias
diameter
35 microns

start with one cell—that fertilized egg from which each of us derived, for example—differentiating into different types that form tissues, which in turn can form sophisticated specialized organs. (In others, a new individual may begin by budding—by no means a lesser feat.) The cells of a multicellular organism also produce substances that hold the individual together and compounds that enable intercellular communication. Chapter five looks at how this all works to build a robust, functioning body.

A matter of life and death

Strange as it may seem at first, it is of equal importance for cells to die as it is for them to live. Imagine if all the cells that had ever lived were still alive. Nature is a constant battleground in which cells fight for dominance or sometimes just for survival. Competition for space or resources is a driving force in evolution, a process that simply would not work without death. Chapter six sets out how cells compete and how they die—including the importance of cell self-destruction and the problems that failure to self-destruct can cause. Then the final chapter considers the most interesting and vital cells of the human body, all of which can trace their origins back to the single cell that heralded the beginning of a new person all those years ago.

Human hair cross section
diameter 100 microns (x10 cells)

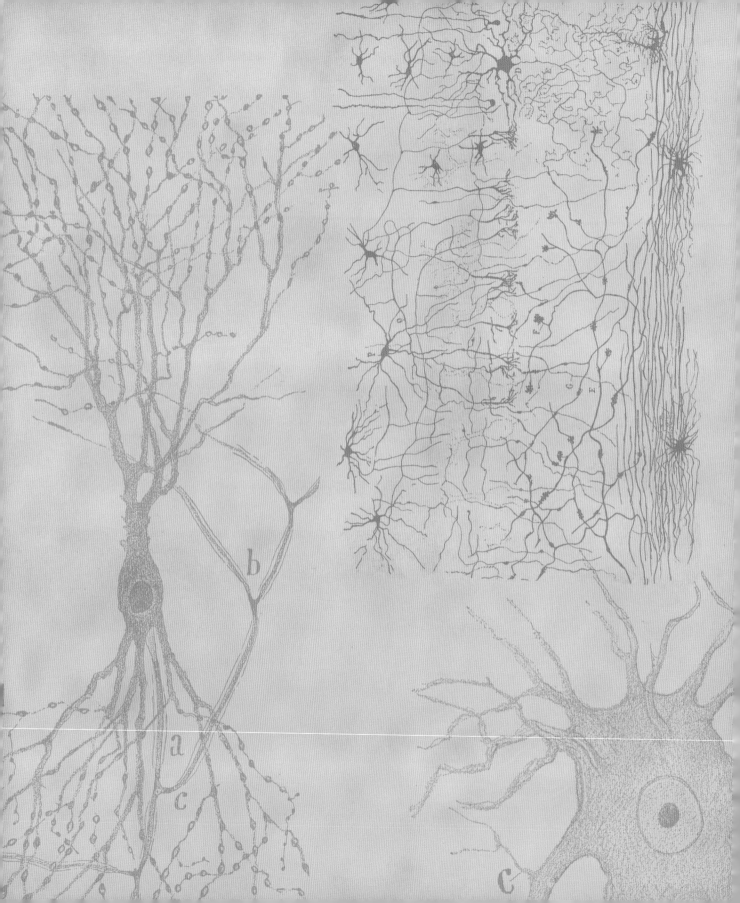

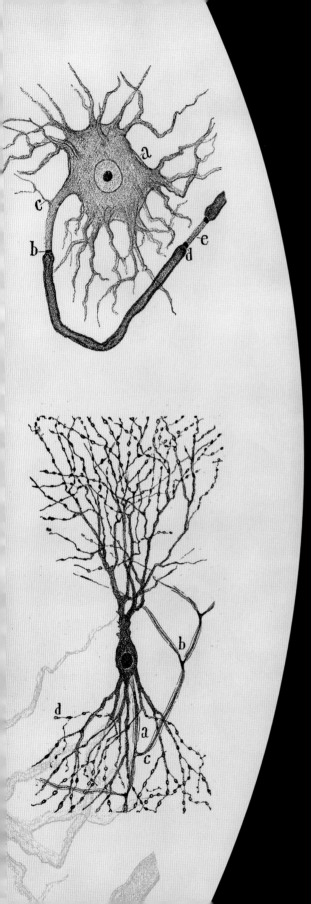

CHAPTER ONE
A Brief History of the Cell

The earliest observations of cells were made in the late seventeenth century, but their fundamental importance in the natural world only became apparent over 150 years later, in the middle of the nineteenth century. Since then, increasingly rapid strides have been taken toward understanding what goes on inside cells—and how such processes relate to growth, reproduction, inheritance, disease, and the origin of life on Earth.

Left *Pioneering Spanish cell biologist Santiago Ramón y Cajal made these remarkable drawings, of interconnected neurons (nerve cells) from the brain of a rabbit, in 1899, just over three decades before the invention of the electron microscope.*

A whole new world

When seventeenth-century natural philosophers and physicians gazed through microscopes at plants, animals and fungi they were treated to tantalizing glimpses of anatomy and physiology on tiny scales. Microscopes allowed these scientists and doctors to discover "microorganisms"—entire living things too small to see with the naked eye—and to stumble across the existence of cells.

A revolution in seeing

The facts surrounding the invention of the microscope are about as clear as the images that early examples of these instruments produced. It was in the 1590s, or possibly the early 1600s, and probably in Holland, but possibly in England, that someone first placed two lenses in an arrangement that produced a magnified image. What is known is that the new instrument, more powerful than the hand lenses already in use, quickly captured the imagination of natural philosophers across Europe.

The magnifying power and optical quality of microscopes improved gradually during the seventeenth century. Although minerals and everyday objects were frequent subjects of study, it was closeup views of living things that really caught people's eyes. In 1660, the Italian physician Marcello Malpighi carried out microscopic studies of human flesh and found tiny blood vessels—the capillaries, which join arteries to veins. The discovery of capillaries confirmed a controversial theory: the circulation of blood, put forward by William Harvey in 1628. Malpighi studied many plants and animals with his microscopes, and in 1666, after studying a

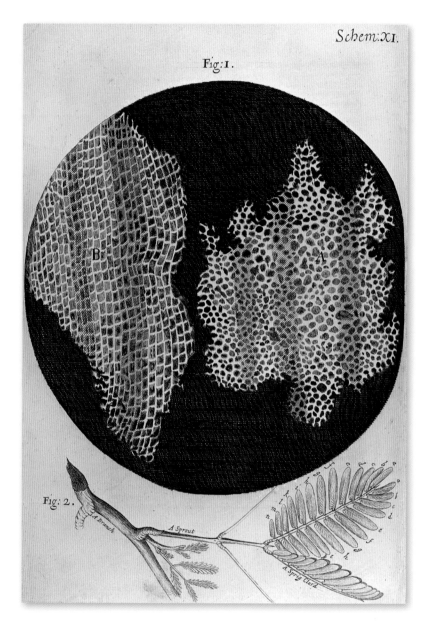

Above *Robert Hooke's drawing of "cells" in cork. What he actually observed were the spaces enclosed by cell walls of empty, dead cells. Note that "B" is, as Hooke described it, "split the long-ways."*

blood clot, he described "very small red particles" that "roll and turn helter-skelter", the first confirmed sighting of what we now call red blood cells.

Tiny boxes

The most influential microscopist of the age was Englishman Robert Hooke. While employed as "curator of experiments" at the new Royal Society in London, Hooke made many observations through microscopes and telescopes, and produced a beautifully illustrated book of what he had seen. *Micrographia* was published in September 1665 and its exquisite drawings and intriguing text gave readers an insight into a world hidden from everyday eyes. The now famous diarist Samuel Pepys was among those captivated, noting: *"Before I went to bed, I sat up till 2 a-clock in my chamber, reading of Mr. Hooke's Microscopical Observations, the most ingenious book that I ever read in my life."*

It was Hooke who coined the word "cell" to describe what he saw when studying cork. He placed thin slivers of the material onto a dark plate beneath his microscope's objective lens, illuminated them with light from an oil lamp focused through a thick lens, and gazed through the eyepiece. His description of what he saw, quoted below, is still intriguing.

Hooke estimated that there were about 10,000 cells to the inch (about 4,000 per centimeter) and that one cubic inch would contain about "twelve hundred millions" (about 70 million per cubic centimeter). It was an astonishing discovery; he wrote that this intricate structure was "almost incredible, did not our Microscope assure us of it by ocular demonstration." Each of Hooke's "cells" is a cube with sides measuring just over 20 microns, or 0.02 millimeters.

> *"I could exceeding plainly perceive it to be all perforated and porous, much like a Honey-comb, but that the pores of it were not regular … these pores, or cells, were not very deep, but consisted of a great many little Boxes, separated out of one continued long pore, by certain Diaphragms, as is visible by the Figure B, which represents a sight of those pores split the long-ways."*

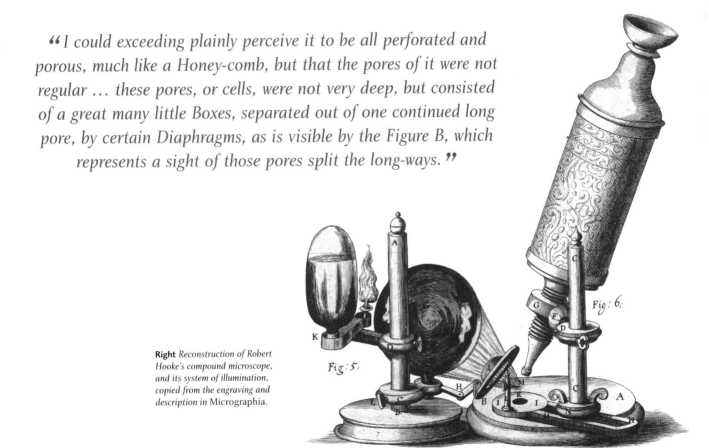

Right *Reconstruction of Robert Hooke's compound microscope, and its system of illumination, copied from the engraving and description in* Micrographia.

A BRIEF HISTORY OF THE CELL

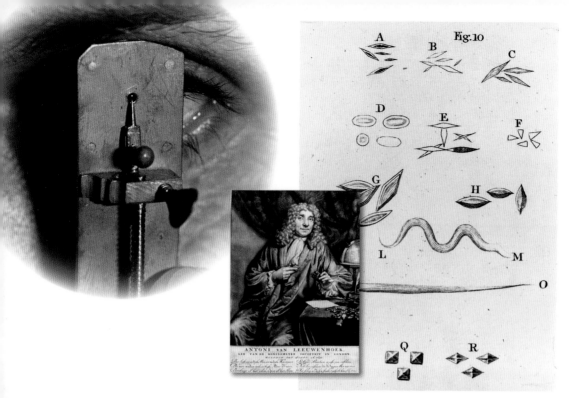

Left *Leeuwenhoek (depicted center) was prolific; he made more than 500 microscopes and wrote hundreds of letters informing scientific societies about his discoveries—including 190 or so to the Royal Society in London. The drawings (right) are taken from one of his letters. Leeuwenhoek's microscope (far left) was a handheld metal frame with screws to adjust the stage (sample holder) and to move the lens for focusing.*

Animalcules

Although compound microscopes (with two or more lenses) were popular during the seventeenth century, many investigators also used "simple microscopes"—just single powerful lenses. Some of these could magnify as well, if not better, than their more complicated counterparts. One man who favored single lenses was Antony van Leeuwenhoek, a successful Dutch draper. Leeuwenhoek made tiny, near-spherical lenses by melting glass rods in a flame. He carefully ground them to the right shape and attached them to ingenious handheld metal frames that also held the specimen. He investigated everything from tongues to sand and became the first person to describe sperm cells (which he found in the males of several species, including humans). While most microscopes of the time achieved magnifications of between 30x and 60x, Leeuwenhoek's could magnify more than 250x.

In 1675, Leeuwenhoek observed tiny living creatures in a sample of rainwater that had been standing for a few days. These microorganisms were far, far smaller than any living things anyone had ever seen. Leeuwenhoek called them "animalcules." For the next year he studied river water, well water, and seawater, some of which he left standing for several days or weeks. Mostly, he saw protozoa and single-celled algae, which are about the same size as Hooke's cork cells—some quite a lot larger. But in April 1676 he saw animalcules that were much smaller, and these he described as being so tiny that you would need to lay more than a hundred end to end to measure the same as a grain of sand. This was almost certainly the first observation of bacteria.

A letter that Leeuwenhoek wrote in 1683 contains the world's first illustrations of bacteria. The letter detailed his microscopic investigations of his own dental plaque: "I have mixed it with clean rain water, in which there were no animalcules, and I saw with great wonder that there were many very little animalcules, very prettily a-moving." Leeuwenhoek also wrote that "there are not living in our United Netherlands so many people as I carry living animals in my mouth this very day."

A BRIEF HISTORY OF THE CELL

1 NORMAL VISION

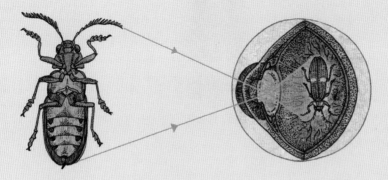

2 WITH MAGNIFYING GLASS

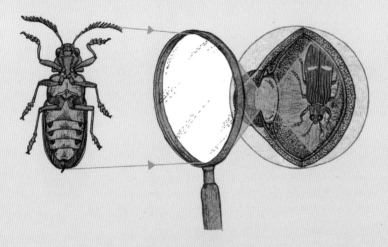

3 OPTICAL MICROSCOPE

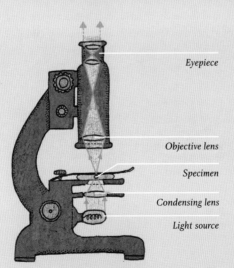

HOW OPTICAL MICROSCOPES WORK

Light passing through or bouncing off any point of an object spreads out in all directions, in straight lines, or rays. Any rays that pass through the eye's lens form an image on the retina (1). All the rays from any particular point of the object always reach the corresponding point of the image, thanks to the focusing ability of the eye's lens. It is possible to illustrate the extent of the image produced on the retina by choosing just two points—one at the top and one at the bottom of the object—and just one ray from each point. The two rays chosen here pass through the center of the lens, without bending. The apparent size of an object is determined by how much of the retina the image takes up, which in turn is determined by the angle at which these two rays enter the eye: the visual angle. Lenses and microscopes change the visual angle, by bending light—and by doing so, they can enlarge the retinal image, making an object appear much bigger (2).

A basic compound light or "optical" microscope (3) consists of a light source, a stage on which to place the specimen and two lenses (or sets of lenses) called the objective lens and the eyepiece. The objective lens, the one closest to the specimen, produces a magnified "real image" of the specimen inside the microscope tube. This simply means that if a piece of paper were placed there, the image would be projected. The eyepiece acts as a magnifying glass, enlarging that first image to produce a very high magnification overall. The total magnification is the magnification of the objective lens multiplied by the magnification of the lens or lenses in the eyepiece.

Cell theory

Surprisingly, perhaps, the idea that living things are made of cells did not come from the observations of the first microscopists, such as Hooke, Leeuwenhoek, and Malpighi. Instead, it originated as a philosophical thought borrowed from physical science.

Living molecules

In his influential book *Philosophiæ Naturalis Principia Mathematica*, published in 1687, the English scientist Isaac Newton popularized the notion that matter and even light might be made of tiny, indestructible particles. This idea had a long pedigree, not least because the alternative is continuous matter, which is difficult to understand. Many naturalists wondered whether living things might be made of particles of a different kind; because they believed living things are fundamentally different from nonliving matter, the particles themselves would have to be alive. In 1749, French naturalist Georges-Louis Leclerc, the Comte de Buffon, called them "living molecules."

At the same time, biologists were busy familiarizing themselves with the microscopic anatomy, or histology, of plants and animals. Some wrote about their observations of "cellular tissue" and even began to make the connection between their observations and the idea of living molecules. However, many of the "cells" were optical illusions caused by dirty lenses or out-of-focus microscopes— and, in plants at least, the word "cellular" often meant "populated by empty spaces." By the early nineteenth century, however, scientists had begun focusing their ideas—and their microscopes— more acutely.

Coming together

In the 1820s French botanist Henri Dutrochet boiled plant tissue in nitric acid to dissolve away the material that held the cells together. He watched as

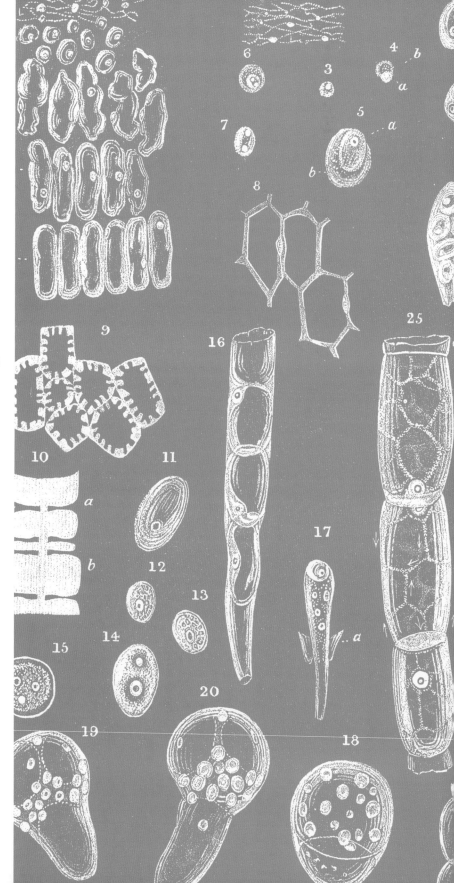

the cells separated into countless individual, self-contained "globules," concluding that cells make up "the fruits, the stems, roots, leaves and flowers on all the plants on the surface of the planet." The German botanist Franz Meyen reached a similar conclusion in 1830, observing: "Plant cells occur either singly, so that each forms an individual, as in the case of some algae … or they are united together to form greater or smaller masses, to constitute a more highly organized plant."

Some botanists observed a dark spot inside certain plant cells. In 1831 British botanist Robert Brown gave it a name: the "nucleus." In the 1830s, another German botanist, Matthias Schleiden, suggested that nuclei were the source of new cells after he supposedly saw new cells forming around nuclei and then emerging from inside the cell. In 1837, Schleiden described what he had seen to his colleague, the zoologist Theodor Schwann, who recognized the description of the dark spots as something he had seen in the cells of tadpoles. Schwann was convinced that in animals, too, nuclei seemed to give rise to new cells.

Two out of three

Schwann thought he also saw nuclei in the spaces between cells. He suggested that all nuclei crystallize from a hypothetical substance which he called "cytoblastema," believing it to be present inside and outside cells. In 1839, he established the first coherent cell theory, based on three principles: first, that every part of every living thing is either made of cells or made by cells; second, that the life of a living organism as a whole is due to the fact that its cells are alive; third, that cells come into existence in or near other cells, from cytoblastema. The last idea was quickly gunned down by biologists who had observed cells dividing in two (binary fission) and realized that they were reproducing.

French biologist Barthélemy Dumortier had watched binary fission as early as 1832, even writing that it provided a "perfectly clear explanation of the origin and development of cells," but his observation had remained controversial. Robert Remak, a Polish-German embryologist, carried out painstaking studies of developing embryos in the 1840s and noticed that every new cell arose from the division of preexisting cells.

In 1855, German biologist Rudolf Virchow made Remak's observations his own and established and publicized modern cell theory. He dropped Schwann's idea about cytoblastema, replacing it with a Latin phrase: *omnis cellula e cellula* ("every cell comes from a cell"). This simple idea is crucial to understanding how living things grow and reproduce and, ultimately, for making sense of the process of evolution by which species rise and fall over time.

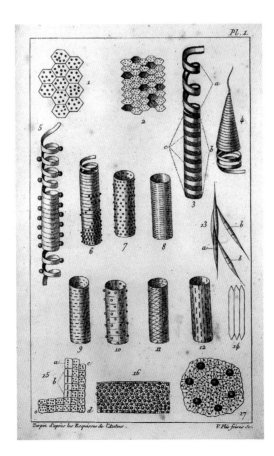

Left *Illustrations by Matthias Schleiden, detailing his observations of plant cells and their nuclei (cytoblasts), from an 1847 book by Theodor Schwann.*

Right *Drawings of plant cells by Henri Dutrochet, 1824.*

A BRIEF HISTORY OF THE CELL

Passing it on

Cell theory, combined with improvements in microscopy, made sense of the living world—in particular growth, reproduction and inheritance. Asking ever deeper questions, cell biologists turned to the chemistry of living things: biochemistry.

Seeing more clearly

Theodor Schwann's idea about cell nuclei crystallizing from cytoblastema is not as absurd as it might seem. The nucleus becomes far less prominent when a cell is about to divide, so it really can seem to appear and disappear. It took two technological advances to show that the nucleus is present all the time and to begin working out its role.

The first of these advances was better microscopes. In the 1830s, British physicist Joseph Jackson Lister introduced microscopes with lenses that corrected for spherical aberration (distortion of the image) and chromatic aberration (annoying colored fringes around the image). Ernst Abbe, a German physicist, pushed optical microscope design to its limits in the 1870s, immersing lenses in oil to maximize magnification, resolution, brightness, and contrast.

The second important advance was histological staining—the use of dyes that are absorbed only by certain structures in the cell to make those structures stand out. Working with brain cells in 1858 German anatomist Joseph von Gerlach noticed how carmine (cochineal) was taken up by the nucleus and its contents but not by the rest of the cell. The introduction of a range of synthetic dyes in the 1860s opened up the technique, leading to the discovery of several other organelles (see box opposite).

Identical twin daughters

The nucleus does not disappear during cell division. Instead, it splits into pieces, which are shared out between the two resulting "daughter" cells. In 1882 German biologist Walther Flemming named the material from which the fragments are made chromatin (because it readily absorbed the colored stain he was using). Working with cells from salamanders, Flemming also noticed that during cell division, the chromatin becomes arranged into distinct strands, later named chromosomes; these are pulled apart by tiny, barely visible ropes to form the nuclei of the two daughter cells. The perfectly coordinated dance of tiny colored strands inside a cell nucleus was (and still is) a wonder to behold. He called it mitosis, a term still used today. An easily overlooked point—but one that reveals a profound unity throughout nature—is that soon after

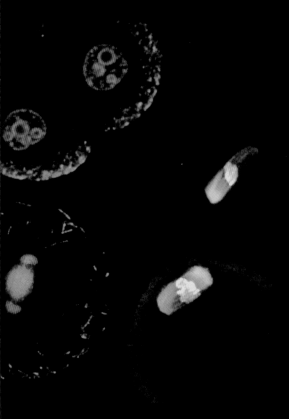

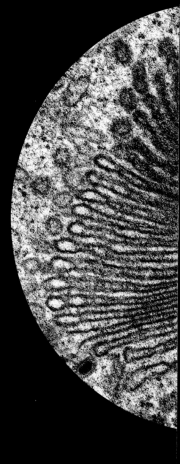

Left *Micrograph showing the dramatic changes that take place during cell division. This modern image uses fluorescent stains to highlight the nuclear material (red) and the protein components of the cell (green).*

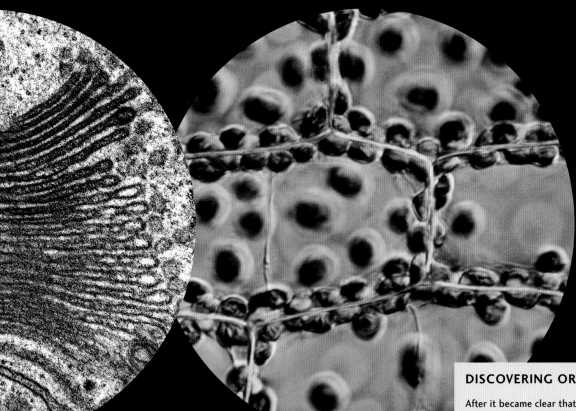

Left *Modern views of two organelles discovered in the 19th century. Far left: A colored transmission electron micrograph of a Golgi apparatus. Left: A phase contrast light micrograph of chloroplasts (the green blobs) in plant cells.*

Flemming observed mitosis in salamander cells, others soon saw the same dance happening in other animals, plant cells, and fungi (bacterial cells do not have nuclei and reproduce differently, as explored in chapter four).

Biologists studying mitosis during cell division noticed that each of the daughter cells ended up not with just a random half of the chromatin material but identical sets of chromosomes. Quite how the cell manages to do that would remain a mystery for decades to come. However, it suggested that the chromosomes carry information essential to the proper functioning of the cell, or perhaps even the whole organism—something like a set of instructions.

DISCOVERING ORGANELLES

After it became clear that there are a number of structures inside cells akin to the organs in the human body, German zoologist Karl Möbius suggested the name "organulas" (little organs). The term for them later became "organelles." The first organelle to be discovered was the nucleus.

Determined to find out how plant cells manufacture starch, German botanist Julius von Sachs discovered the chloroplast (a structure essential in photosynthesis) in the 1860s—although it wasn't named until the 1880s. Several biologists had reported seeing granules in muscle cells from the 1840s onward; they were identified in other types of cell in the 1890s, and called mitochondria. The "Golgi apparatus" was discovered by Italian physician Camillo Golgi during an investigation of the nervous system in 1897, although no one understood its function for decades to come. Another large, if inconspicuous, organelle is the endoplasmic reticulum. It was discovered in 1902 by Italian pathologist Emilio Veratti.

A BRIEF HISTORY OF THE CELL

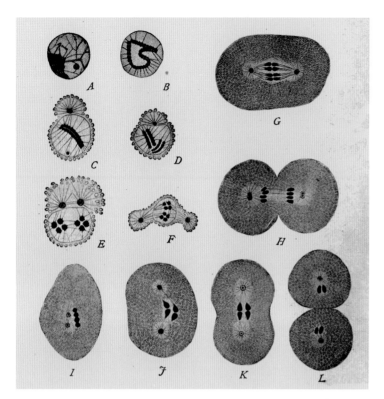

Left *Early observations of meiosis during the production of sperm cells in the worm. Drawings in* The Text-book of Embryology *by Fredrick Bailey and Adam Miller, first published in 1909.*

Above *Gregor Mendel, the Austrian monk who founded the science of genetics with his plant-breeding experiments in the 1860s. The importance of his findings went unrecognized for over three decades.*

Sex cells

The nature of the information carried by chromosomes became a little clearer in the last few years of the nineteenth century. Biologists noticed that the nuclei of egg and sperm cells (sex cells, or gametes) have half as many chromosomes as cells in ordinary tissue (somatic cells). The process by which the set of chromosomes in a nucleus is reduced by half is another remarkable and intricate chemical dance that came to be called meiosis.

During fertilization, the two nuclei fuse together to form a complete, and new, set of chromosomes, different from the chromosome set of either the male or female parent. In that way, both parents contribute equally to the chromosome set of the new individual created by fertilization. It was clear that the chromosomes have something to do with heredity—passing characteristics from generation to generation. They provide instructions for cell housekeeping and a set of instructions on how to build a unique organism, both contained in one tiny part of each tiny cell.

In the first few years of the twentieth century, biologists studying heredity rediscovered the work of an Austrian monk, Gregor Mendel. In the 1860s, Mendel had conducted experiments on pea plants, trying to work out the rules of inheritance. He established that each physical trait is carried by factors that come in pairs, one from each parent. Where an individual inherits two different forms of a particular factor, one form is usually dominant. So, for example, if a purebred pea plant with yellow seeds is crossed with a purebred plant with green seeds, the offspring will all have yellow seeds. But cross two plants from that generation, and some will have green seeds, because they have retained the nondominant (or recessive) factor and passed it on to their offspring. Only if an individual plant inherits two copies of the recessive green-seed factor will it have green seeds.

A new generation

German biologist Theodor Boveri and American biologist Walter Sutton realized that Mendel's factors are somehow carried by the chromosomes—one from each parent—and that Mendel's laws of inheritance are applied inside the cell nucleus. In 1909, Danish botanist Wilhelm Johannsen came up with a name for this bold and exciting new science: genetics. Mendel's "factors," carried on chromosomes, became genes. In a laboratory in New York, a team of scientists led by geneticist Thomas Hunt Morgan began the process of mapping certain physical characteristics to certain chromosomes, and even to certain regions of particular chromosomes.

Science is a never-ending journey toward the truth; any answer to one question opens up a deeper layer of questions. In this case, a new line of investigation was: "What is chromatin, and how does it carry information?" The answer would begin with chemistry—biochemistry—and would end up spawning a whole new discipline: molecular biology.

Breaking it down

Because a cell is, in essence, nothing more than a bag of chemicals, it is vital for anyone trying to understand the cell to determine its contents. In the nineteenth century, biochemists had worked out that chloroplasts in plant cells manufacture glucose and starch. These compounds are carbohydrates that are used as a source of energy; they are also built up into larger molecules of cellulose (the main structural material of plants). The scientists also knew that cells contain fats—in the cell's outer envelope, or membrane, and inside the cell, too. And they were aware that cells manufacture proteins, which are the main component of most biological materials. Hair and nails are made almost exclusively of proteins, for example, and skin owes its toughness and elasticity to two proteins, collagen and elastin.

In the 1870s biologists also began discovering enzymes—catalysts that encourage chemical reactions to take place. The fermentation of sugar, for example, does not happen spontaneously, but it does proceed in the presence of an enzyme called zymase. The nature of enzymes was a mystery until the painstaking experiments of American chemist James Sumner in the 1920s proved that enzymes are, in fact, proteins. So ubiquitous are proteins in the cell that many biologists believed that proteins had to be the carriers of heredity. But they were wrong.

Below *Purebred green (gg) and yellow (YY) pea plants can only produce yellow peas in the first generation, because the yellow allele is dominant. But in the second generation, one in four plants will have green peas (gg).*

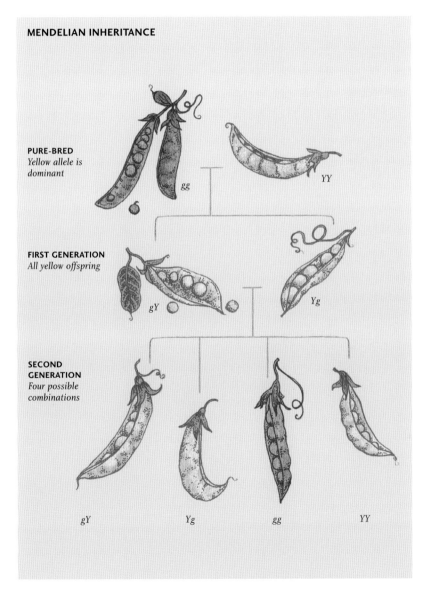

MENDELIAN INHERITANCE

PURE-BRED
Yellow allele is dominant

gg   YY

FIRST GENERATION
All yellow offspring

gY   Yg

SECOND GENERATION
Four possible combinations

gY   Yg   gg   YY

Intimate knowledge

During the twentieth century the mysteries of reproduction and inheritance finally succumbed to science's relentless quest for understanding. Biochemistry and molecular biology have laid bare the machinery of life at the smallest scale. Meanwhile, electron microscopes have given cell biologists access to ever clearer views of cells' components, and a range of new techniques for looking inside living cells has provided intimate knowledge of those components at work.

Nuclear chemistry

While most biologists supposed that proteins must be the carriers of genetic information, another candidate was waiting in the wings. In 1869, Friedrich Miescher, a Swiss biochemist, isolated a mysterious compound from inside the nuclei of white blood cells. Because he discovered it in the nucleus, he called it "nuclein." In the 1870s, a German biochemist, Albrecht Kossel, showed that nuclein is made of proteins plus another compound. That other component soon gained the name "nucleic acid" (because it was found to contain phosphate, which is acidic when dissolved in water). In addition to phosphates, nucleic acids were also found to contain sugars. By the turn of the century, Kossel had successfully established that nucleic acids also harbor five compounds: adenine (A), cytosine (C), guanine (G), thymine (T), and uracil (U). These compounds are bases (they form alkaline solutions in water) and so became known as nucleobases.

Biochemists had identified two different types of nucleic acid. They initially called these yeast nucleic acid and pancreas nucleic acid, because of where they were discovered. During the 1910s and 1920s, the Lithuanian-American biochemist Phoebus Levene identified the nature of the sugar molecules present in nucleic acids. Yeast nucleic acid contains a type of sugar called ribose, and so gained the name ribonucleic acid (RNA). Pancreas nucleic acid contains a very similar sugar, called deoxyribose; it gained the name deoxyribonucleic acid (DNA). Levene also worked out that RNA contains only the nucleobases C, G, A, and U, whereas DNA contains only C, G, A, and T, and he came up with a largely unconvincing way in which nucleic acids might be the carriers of hereditary information.

Transforming opinion

Levene's idea was wrong, but he was looking in the right place. Working out the structure of nucleic acids—how the phosphates, sugars, and nucleobases fit together—would ultimately unravel the secrets of how cells make proteins and how chromosomes carry genetic information. Despite the interest in DNA and RNA, most molecular biologists of the 1920s and 1930s clung to the idea that inheritance was carried by proteins.

That view changed after scientists studying bacteria succeeded in transforming one bacterial species into another by mixing dead bacteria of one pathogenic (disease-causing) species with live bacteria of a benign (meaning harmless) one. Something had transferred from the virulent cells to the harmless ones—something species-defining. A British bacteriologist, Frederick Griffiths, was first to do this, in 1928. But it took until 1944 for molecular biologists to establish that what Griffiths had called the "transforming principle" was in fact the nucleic acid DNA.

In what proved to be one of the most important and well-known breakthroughs in the history of science, in 953 scientists in Cambridge and London, England, worked out the structure of the DNA molecule, and thus gave rise to modern molecular biology.

Above *Portrait of Friedrich Miescher, who discovered nuclein, the compound containing DNA, in pus obtained from a hospital adjacent to his laboratory.*

Right *James Watson (seated) and Francis Crick with their model of the double-helix structure of DNA. Metal plates and rods represent nucleobases, sugars, and phosphate groups, and are held in the right place by laboratory clamps and a retort stand.*

Above *Biochemist Phoebus Levene, who discovered the ribose and deoxyribose sugars in nucleic acids.*

Above *Rosalind Franklin, whose work in X-ray crystallography helped determine the structure of DNA. She discovered the double helix independently of Watson and Crick.*

Right *"Photo 51" (taken by Franklin's Ph.D student Raymond Gosling). This image was crucial in determining the positions of the bases, sugars, and phosphates in DNA.*

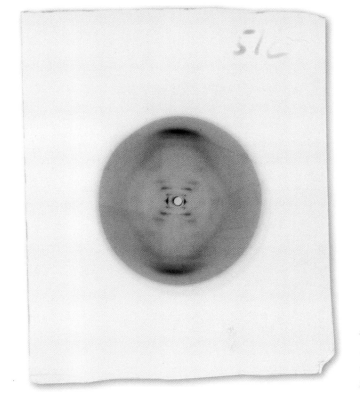

Doubling up

In London, two molecular biologists, Rosalind Franklin and Maurice Wilkins, used a technique called X-ray crystallography to work out the relative positions of the atoms present in DNA. In Cambridge, James Watson and Francis Crick used the X-ray crystallography data to build a model of the DNA molecule made of hundreds of metal plates that represented the sugars, phosphates, and nucleobases. Watson and Crick's model revealed the fact that DNA must have a double-helix structure—something like a twisted rope ladder.

The nucleobases form the ladder's "rungs"; they are strung along two sugar-phosphate backbones that form the ropes, and they join together in the center, preventing the two ropes from coming apart. It is the sequence of nucleobases along the backbones that carries genetic information—forming the letters of the code of life. The letters spell out recipes for creating proteins. In general, each recipe, encoded in a finite sequence of nucleobases, represents a single gene.

Cracking the code

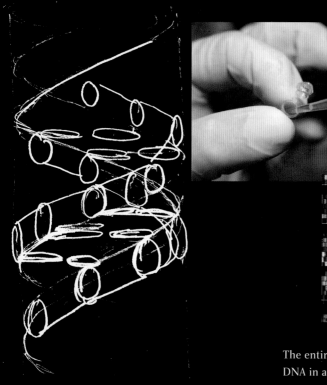

Far left *The pencil sketch of the DNA double helix by Francis Crick.*

Left *A researcher removes DNA (in solution) from a plastic micro test tube using a precise micropipette, ready for sequencing.*

Below *The result of DNA sequencing—a small section of a genome with each of the four nucleobases represented by a different color.*

By the mid-1960s, molecular biologists had worked out the code by which sequences of As, Cs, Ts, and Gs of the nucleobases represent the instructions to make proteins – and the role of RNA in the whole process. When a protein is to be made inside a cell, the two 'ropes' of the DNA double-helix ladder unzip, and a strip of RNA is made as a copy of the code for the relevant gene. The piece of RNA is used as a template for building the protein for which the gene codes. Yes, it's another intricate, well-choreographed molecular dance—but, unlike mitosis and meiosis, it's going on every second of every day, not just when a cell divides. The actual protein-building part is carried out by molecular machines called ribosomes, which were discovered in 1954 by Romanian-American cell biologist George Palade. We will explore clearly and in more detail how this remarkable scheme works in chapter two.

The entire sequence of nucleobases—across all the DNA in all the chromosomes inside the nucleus of an organism's cells—is known as that organism's genome. From the 1970s onward molecular biologists began developing techniques for sequencing DNA—reading the sequence of nucleobases. Soon they were sequencing entire chromosomes and eventually entire genomes. The first genome to be sequenced, in 1995, belongs to a bacterium. A draft version of the three billion-nucleobase-long human genome was completed in 2000, and the Human Genome Project was declared complete in 2003. Of course, everyone's genome is unique, so the project will never really be complete.

Molecular biology, together with genetics, have made remarkable strides in advancing our understanding of what makes living things alive—and how cells manufacture proteins and how information is passed from parents to offspring. Because every cell comes from another cell, and

Right *The building blocks of proteins—amino acids—have been detected in outer space, and these crucial ingredients for life may have been delivered to Earth by comets crashing into our planet during its formative years.*

DNA may be slightly altered as it passes from generation to generation, molecular biology is important not only in making sense of living things today but also in answering questions about the history of life on Earth—evolution—and even its origins more than three billion years ago.

Getting a reaction

There is more to biochemistry than the nucleic acids that carry genetic information – and there is also more to biochemistry than knowing what chemical compounds are present in a cell. It is also important to work out how fats, carbohydrates, and proteins interact with each other and how they are broken down and built up. The set of chemical reactions that keep living things alive is called metabolism.

Two of the most important metabolic reactions are photosynthesis and cellular respiration—although they are actually pathways consisting of many separate, successive reactions. These two reactions are essentially the reverse of one another. It is through photosynthesis that the cells of plants and certain other organisms capture energy from sunlight, storing it in the form of carbohydrates. Cellular respiration releases that energy.

Finding the energy

Working out the many complex compounds and reactions involved in photosynthesis and cellular respiration were mammoth tasks, achieved in the 1930s and 1940s. Both reactions involve a remarkable compound called adenosine triphosphate (ATP), which acts as the cell's energy currency. In photosynthesis, ATP is one of the end products; in cellular respiration, it is both one of the initial reactants and one of the end products. German biochemist Karl Lohmann discovered ATP in 1929, but the manner in which it stores and releases energy was not uncovered until the 1940s.

ATP is built up and broken down by an enzyme called ATP synthase (enzymes always end with the suffix "-ase"). This extraordinary molecule is the world's smallest motor: one part of the molecule literally rotates continuously inside the other part. It took decades for biochemists to work out the structure and behavior of this molecule. Every one of the 50 trillion or so of your living cells

---

THE ORIGIN OF LIFE

No one knows exactly when or how the first cells came to be, but some of the key molecules of life occur naturally in a number of lifeless locations. For example, nucleobases of extraterrestrial origin have been found in meteorites, and amino acids, the building blocks of proteins, have been detected in comets. Another possibility is that the first self-replicating organic molecules were manufactured in the tumultuous conditions of hydrothermal vents—cracks in Earth's crust at the bottom of the ocean. Either way, the world's first cells would probably have been simple oily blobs that contained tiny chemistry laboratories, protecting early self-replicating RNA and DNA from an otherwise harsh environment.

Clearer views

Despite the improvements in microscope technology in the nineteenth century, some features of the cell remained unseen. This is because of theoretical limits to the magnification and resolution possible with optical microscopes. Light waves change direction, or diffract, as they pass around edges or small objects—in the way that waves in water change direction as they pass around the entrance to a harbor. Diffracted light waves from two edges or two objects close together interfere (just as the spreading waves from two adjacent openings in a harbor wall would create choppy water). This interference compromises the quality of the image you can achieve, to the point where you can discern no detail whatsoever. However good the microscope, it is thought to be impossible to overcome this interference. Ernst Abbe, the German physicist who improved microscope design in the 1870s, worked through the math and found that optical microscopes would never be able to resolve two objects separated by less than half the wavelength of light – about 200 nanometers (0.0002 millimeters). Several significant organelles and other features inside the cell are a lot smaller than that.

In 1931, two Germans, the physicist Ernst Ruska and the engineer Max Knoll, invented a new kind of microscope, which uses electrons instead of light to produce images. In 1933, an electron microscope exceeded the theoretical magnification of optical microscopes. By the late 1930s, electron microscopes could resolve objects 10 nanometers apart—by the mid-1940s, 2 nanometers. One of the first triumphs of the electron microscope in cell biology was the discovery of ribosomes—the molecular machines that build proteins.

New vision

Cell biologists now have at their disposal a variety of innovative techniques to use in conjunction with optical and electron microscopes. These provide scientists with unprecedented views of the inner life of the cell. Dark-field microscopy, for example,

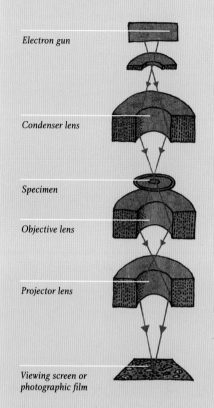

TRANSMISSION ELECTRON MICROSCOPY (TEM)

- Electron gun
- Condenser lens
- Specimen
- Objective lens
- Projector lens
- Viewing screen or photographic film

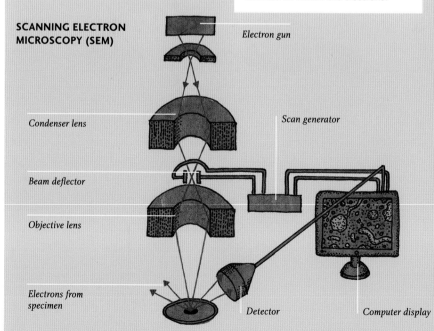

SCANNING ELECTRON MICROSCOPY (SEM)

- Electron gun
- Condenser lens
- Beam deflector
- Objective lens
- Electrons from specimen
- Scan generator
- Detector
- Computer display

HOW ELECTRON MICROSCOPES WORK

The discoveries of quantum physics in the first 30 years of the twentieth century led to the inescapable and bizarre conclusion that electrons, discovered in 1898 as tiny charged particles, are, in fact, waves as much as they are particles. Similarly, light is a stream of particles, called photons, as much as it is a wave. In each case, the beams interact with objects, because they either bounce off or pass through, and if you have a lens, you can form an image of those objects (the lenses in electron microscopes are electromagnets, not glass). Because the wavelength of electrons is so very much smaller than the wavelength of light, an electron microscope can resolve objects and other details of a much smaller size than an optical microscope can.

In a transmission electron microscope (TEM), electrons pass through a slice of the specimen. In a scanning electron microscope (SEM), the specimen is given a very thin coating of metal to reflect the electrons.

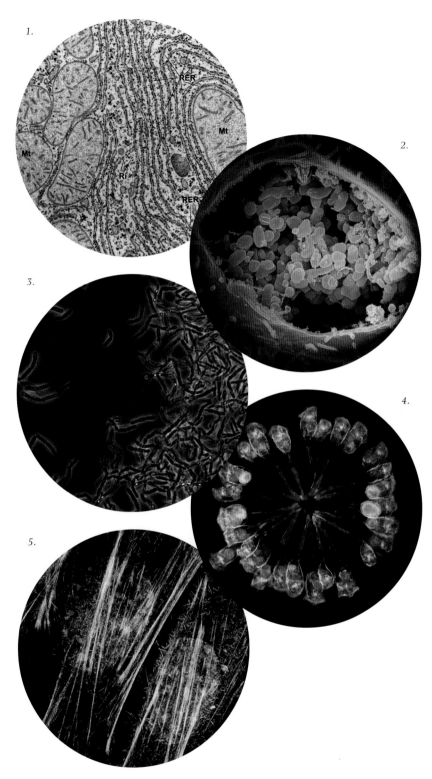

invented in 1909, is a way of making everything except the specimen appear black. Phase contrast microscopy, invented in the 1930s, provides a way of picking out details in otherwise translucent organelles without the use of staining.

Perhaps the most exciting recent innovation is the development of fluorescence microscopy. Sometimes this takes advantage of a specimen's natural fluorescence—for example, particular proteins produce visible light when illuminated with ultraviolet radiation (UV). Alternatively, a sample can be stained with fluorescent dyes. Most exciting is when an organism's DNA is deliberately altered to add a sequence that codes for a fluorescent protein. When that part of DNA is active, the protein is also made, and the activity can be followed in a living cell in real time. The fluorescent protein most frequently used is green fluorescent protein, discovered in jellyfish in the 1960s and incorporated into a wide range of plants and animals by genetic engineering since the 1990s.

Throughout this book are remarkable images produced by optical microscopes and electron microscopes, using phase contrast, dark-field illumination, and fluorescence. Robert Hooke and Antony van Leeuwenhoek would have been amazed.

1. Transmission electron micrograph showing mitochondria and endoplasmic reticulum inside a cell. Like all electron micrographs, there is no color.

2. Scanning electron micrograph parasitic bacteria living inside a larger cell. "False color" has been added to differentiate the different elements of the image.

3. Phase contrast light micrograph showing the bacterium Bacillus anthracis. Dry spores inside some of the bacterial cells show up as bright dots, because they refract the light differently than the rest of the cell.

4. Dark-field micrograph of "stentor" protozoa. Due to a clever optical arrangement, all light is excluded from the image except light that has been scattered by the object. That explains why the background is dark.

5. Very high magnification fluorescent micrograph of DNA (blue) and protein cytoskeleton filaments (yellow) inside a cell. The DNA and proteins are stained with different fluorescent dyes.

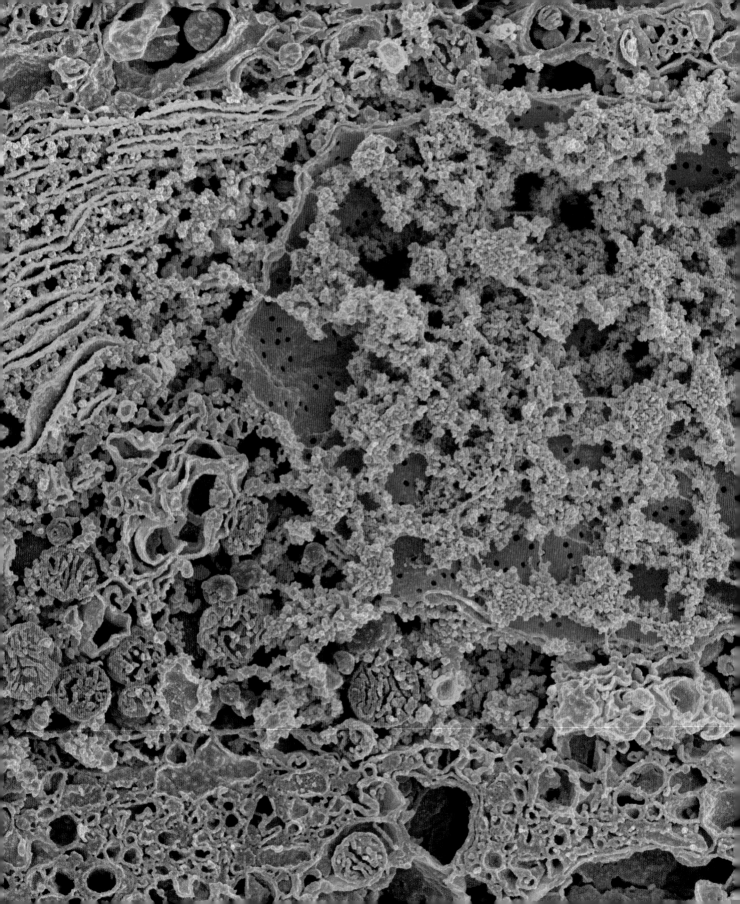

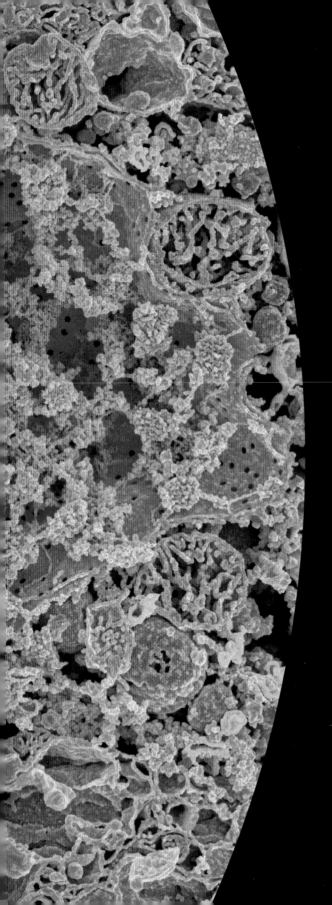

CHAPTER TWO

Inside Living Cells

Over more than three billion years of evolution has honed cells into complex, efficient machines. Science has gazed inside those machines, cataloged each functioning part, and figured out exactly how the processes that sustain life actually work. There is tremendous diversity in the cellular world—but there are certain things that all cells have in common.

Left *False color scanning electron micrograph of part of a single cell—a neuron in the brain of a mouse embryo. DNA is colored pink; the nuclear membrane is purple. Note the pores in the membrane: the endoplasmic reticulum is bright green; ribosomes, which manufacture proteins, are blue; Golgi complexes are colored dull green; mitochondria—the cell's power plants—are red. This remarkable image was produced by removing water from the sample and freezing the sample in liquid nitrogen then breaking it with a razor blade.*

Three domains, two types of cell

All living things on Earth fall into one of three main categories, or domains: Eukaryota, Bacteria, and Archaea. Eukaryota includes a wide range of organisms from single-celled amoeba and yeast to all plants and animals; "eukaryotic" cells are large and complex. Bacteria and Archaea are all simpler, single-celled organisms. Although there are some differences between them, their cells share certain features and are all classed as "prokaryotic."

Eukaryotic and prokaryotic

The domain Eukaryota includes all animals, plants and fungi; humans are eukaryotes, as are falcons, chimpanzees, hydrangeas, and mushrooms. Eukaryota also includes some much smaller things, such as yeasts, amoebas, and algae. What unites this extraordinarily varied collection of organisms is their eukaryotic cells; all share certain features that differentiate them from the prokaryotic cells of Bacteria and Archaea.

The defining characteristic of eukaryotic cells is the presence of a nucleus. Every cell on Earth utilizes DNA to carry its genetic material; in eukaryotic cells, the DNA is housed inside a nucleus. Prokaryotic cells have no nucleus—their DNA is loose inside the cell. There are several other differences. For example, while every cell is enveloped in a double-layer fatty membrane, eukaryotic cells also have a sophisticated system of membranes inside. The internal membranes provide folded surfaces that participate in many of the cell's functions and include the magnificently named Golgi apparatus and endoplasmic reticulum. There are also self-contained organelles, such as mitochondria and chloroplasts, with their own membranes. Prokaryotic cells lack these membrane-bound organelles and membranous structures.

One good analogy to help distinguish between the two types of cell is that a prokaryotic cell is like a village: perfectly complete and functioning but small and relatively simple. A eukaryotic cell, on the other hand, is like a city: larger and more complex, with more structures and more levels of organization. The analogy carries through to comparisons of size, too; eukaryotic cells are typically ten times the size of prokaryotic cells.

Compare and contrast

There are certain things—in addition to using DNA as the carrier of genetic information—that are common to both prokaryotic and eukaryotic cells. For example, all cells use the compound RNA to make copies of sections of DNA (genes), as templates for

Below *All prokaryotes are single-celled organisms. Eukaryotes generally have larger cells, because they have more components, or organelles. Most eukaryotic organisms are multicellular, but there are also single-celled eukaryotes, such as yeasts and many algae.*

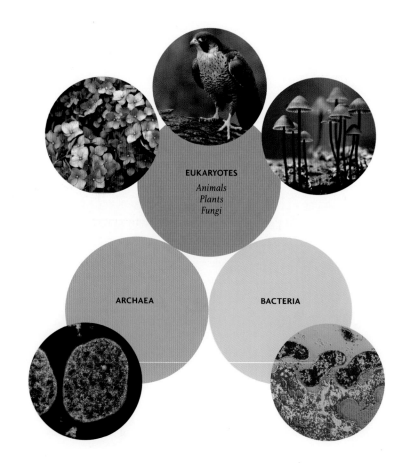

INSIDE LIVING CELLS

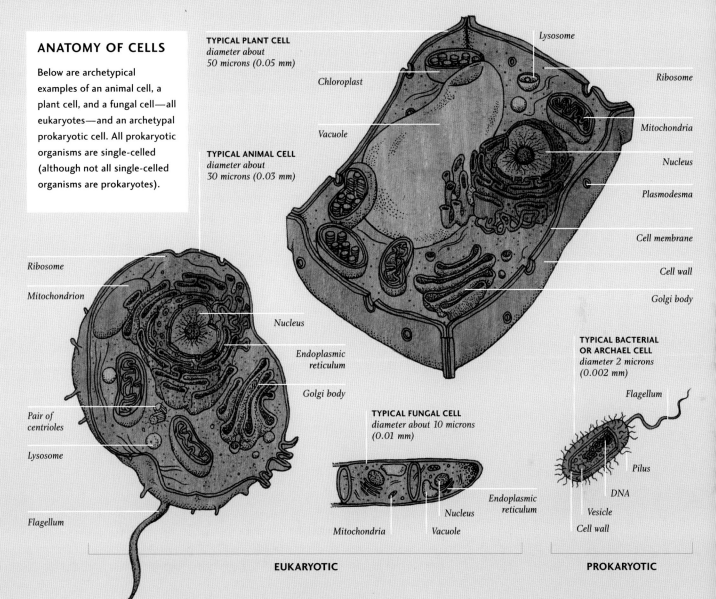

ANATOMY OF CELLS

Below are archetypical examples of an animal cell, a plant cell, and a fungal cell—all eukaryotes—and an archetypal prokaryotic cell. All prokaryotic organisms are single-celled (although not all single-celled organisms are prokaryotes).

building proteins. And in both types of cell molecular machines called ribosomes "read" the RNA and actually construct the proteins. (The ribosomes in eukaryotic cells are slightly different from those in prokaryotic cells but the process is very similar.)

All cells also make a variety of motor proteins, which carry out a range of different tasks. In both types of cell, motor proteins might be found in flagella—the whipping tails that help waterborne, single-celled organisms get around. However, there is one important function of motor proteins that is found only in eukaryotes. Kinesin and dynein are remarkable motor proteins that drag other molecules or entire organelles around inside the cell, literally walking along tubular tracks called microtubules (see page 50)—for example, taking proteins manufactured by the cell up to the cell membrane to be released outside the cell. The microtubules are part of a dynamic system of ropes, poles, and tough fibers called the cytoskeleton, which is also found only in eukaryotic cells.

Around the outside

If we were able to shrink down to the size of a typical cell and to peer inside a selection of different cells, what we would see as the outermost part of the cell would depend upon the kind of cell. All cells are enveloped by a translucent, fatty membrane, but with some cells, there are several layers of other material outside that.

Peeling back the layers

Bacteria are among the most protected of all cells. This might come as depressing news to anyone suffering from a bacterial infection—but fortunately, in most cases, the body is quite adept at killing the harmful ones anyway. On the outermost layer, the majority of bacterial cells have a capsule made of a selection of polymers. Polymers are compounds with relatively large molecules made up of repeating, interconnected smaller molecules. In the case of the bacterial cell capsule, the polymers are polysaccharides and the smaller molecules are sugars. The capsule can form a tough coat or a slimy coating; either way, it helps protect the bacterium from being consumed by other cells and helps it stick to surfaces (such as teeth, for example). Protruding out through the capsule are tiny protein fibers called fimbriae, which are also involved in making bacteria stick to surfaces.

Some bacteria have an outer membrane (in addition to the main cell membrane) directly underneath the capsule layer. This outer membrane is present on "gram-negative" bacteria (see box). It provides extra protection against the environment in which the bacterium lives but also has proteins embedded in it that allow only certain chemical compounds in and out. It keeps out harmful substances but allows in nutrients, for example. This selective permeability is something we will encounter again. If the bacterium we are examining has an outer membrane, by removing it we would come across a tough cell wall. Gram-negative bacteria have only a very thin cell wall—after all, they have that outer membrane. Gram-positive bacteria have a much thicker wall.

Bacterial cell walls are made of another polymer, called peptidoglycan. This compound's long molecules are joined together by cross-links, making a very resilient structure. The links are constantly breaking and reforming; certain antibiotic medicines, including penicillin, work by preventing the formation of the cross-links, so that the wall quickly disintegrates.

Break down the wall

Once the bacterium's capsule and outer membrane has been stripped away to reveal the cell wall, we can begin comparing bacterial cells with some other types of cell. Archaeal cells, fungal cells, and algal cells also have cell walls, and so do plant cells—but animal cells do not. The walls of archaeal cells are made of protein. The walls of fungal cells are made

Below *Light micrograph of gram-positive bacteria (purplish), which have a thick cell wall, and gram-negative (pinkish) bacteria, which have a thinner cell wall.*

GRAM-POSITIVE, GRAM-NEGATIVE

In 1884, German microbiologist Hans Gram invented a stain that would help show up bacteria. He found that bacterial cells fall into two different classes. Gram-negative bacteria, with a thin cell wall, stain pink; familiar examples are *E. coli* and *Salmonella*. Gram-positive bacteria, with a thicker cell wall, stain purple; examples are *Bacillus anthracis*, which causes anthrax, and the infamous methicillin-resistant *Staphylococcus aureus* (MRSA).

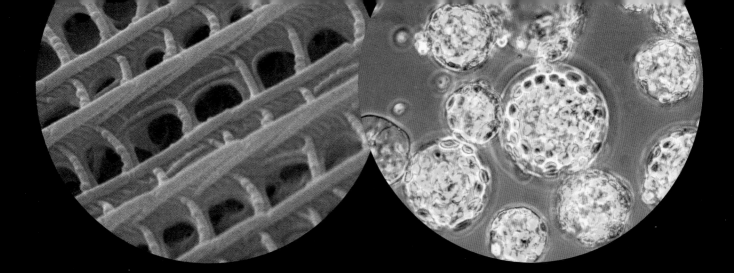

Above left Scanning electron micrograph showing intricate detail of a moth's wing, which is made of chitin, the same material that forms the cell walls of fungi.

Above right Light micrograph of plant cells with their cell walls removed—spherical globules called protoplasts.

of a polymer called chitin. This polymer is found in all kinds of other places in the natural world, including a crab's shell, a squid's beak, and the external skeleton of a housefly.

Algal cell walls come in many flavors, as varied as algae themselves. Algae make up a diverse group of plant-like eukaryotes—everything from single-celled organisms floating free in lakes to seaweeds 54 yards (50 meters) in length. Some algal cell walls are made of proteins, while others are made of polymers of sugars, or "amino-sugars." Some single-celled algae use minerals, such as silicon and calcium, to build an intricate, protective shell. There is more about those elaborate outer casings in chapter four.

The walls of plant cells have an outer covering made of a polymer called pectin (the substance used as a gelling agent in preserves). Pectin helps glue one cell to the next. Below that sticky layer is cellulose, another sugar-base polymer. Cellulose is indigestible to humans. It is the insoluble fiber we ingest from eating plants—and it is those fibers that are visible when a sheet of paper is torn. Some plant cells—those of woody plants, such as trees—also lay down a second cell wall inside the first one. This secondary wall is made mostly of lignin, a still tougher polymer.

Plants' cell walls can remain long after the living cell inside has died. It was the thick cell walls of dead, empty cork cells that Robert Hooke saw down his microscope, when he came up with the name "cells" in 1663. A plant cell's wall prevents the cell from bursting apart when it takes on water, and helps keep its structure when water is scarce.

Naked cells

Remove the cell walls from the bacterial, archaeal, fungal, algal, and plant cells … and we have naked cells. All that is left holding in the watery mixture (the cytoplasm) is a very thin fatty bag (the membrane). The scientific name for a naked cell is a protoplast. Since no animal cells have a cell wall, we can finally include them.

With no capsule, no secondary membrane, and no cell wall, on the face of it animal cells might seem vulnerable. But animals are multicellular, and the material that fills the space between them and holds them together—the extracellular matrix—offers them protection. Cells on the outer surface of the organism, exposed to the world at large, tend to have greater protection, often by secreting tough substances. Think of fish scales or bird feathers. The skin of many animals, including humans, is made of dead, tough, protein-rich cells that form a waterproof, infection-proof barrier. In the case of humans this layer is constantly replenished from below, and tens of thousands of old cells fall off the top layer every minute.

A fluid mosaic

So now we have stripped-down bacterial, algal, fungal, and plant cells—and already naked animal cells. The cell membrane that surrounds these cells is a remarkable structure. Picture an ancient Roman mosaic floor: a pattern of small white tiles with some differently colored tiles interspersed here and there. Now, imagine the tiles shifting around within the floor, in constant motion, and that gives a pretty good idea of what a cell membrane is like. The white tiles represent molecules of phospholipids, the predominant compounds in the membrane. Mixed in are some off-white tiles, which represent molecules of cholesterol. Having cholesterol in the mix makes the membrane more robust.

Each phospholipid molecule has two long lipid tails—the same tails are present in molecules of fat—and a round head containing a phosphate group (see the box below). As might be expected, the fatty tails do not mix with water but the phosphate head does. Put phospholipid molecules in water and they spontaneously arrange themselves tail-to-tail, either in small globules or in double-layered sheets. In the sheets, the phosphate heads point away from each other, making contact with the water. The tails come together, away from the water, staying firmly between the two layers. The sheets have no edges—they wrap around and enclose whatever is inside them, like fluid-filled bubbles.

Two-way traffic

Although the phospholipid bilayer membrane has to keep the cell contents confined, it is by no means impenetrable. If it were, the cell would take in no nutrients and would be unable to get rid of any waste—it would soon die. However, the phospholipid bilayer membrane cannot just let anything in and out; it needs to be selective about what gets through. Water is essential—although it is important that not too much or too little passes through, for obvious reasons. Water can cross the membrane directly but most of it passes—in molecular single file—through tiny holes in ring-shaped protein molecules. These proteins, called aquaporins, straddle the two layers of the membrane (in the Roman mosaic analogy, they are among the colored tiles, interspersed among the white tiles).

Most cells are aerobic. They depend upon oxygen for the energy-releasing cellular respiration reactions that keep the cell ticking over (cells that do not rely

Right *The cell membrane is about 10 nanometers thick (0.00001 mm) but is clearly visible in this coloured transmission electron micrograph.*

Below *Illustration showing a phospholipid molecule (far left) and a small section of a cell membrane, showing how phospholipid molecules self-arrange into a sheet, with their hydrophobic (water-hating) tails pointing away from the watery inside and outside of the cell.*

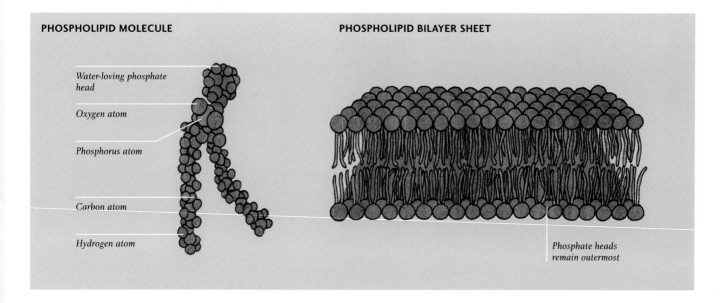

PHOSPHOLIPID MOLECULE

- Water-loving phosphate head
- Oxygen atom
- Phosphorus atom
- Carbon atom
- Hydrogen atom

PHOSPHOLIPID BILAYER SHEET

- Phosphate heads remain outermost

INSIDE LIVING CELLS

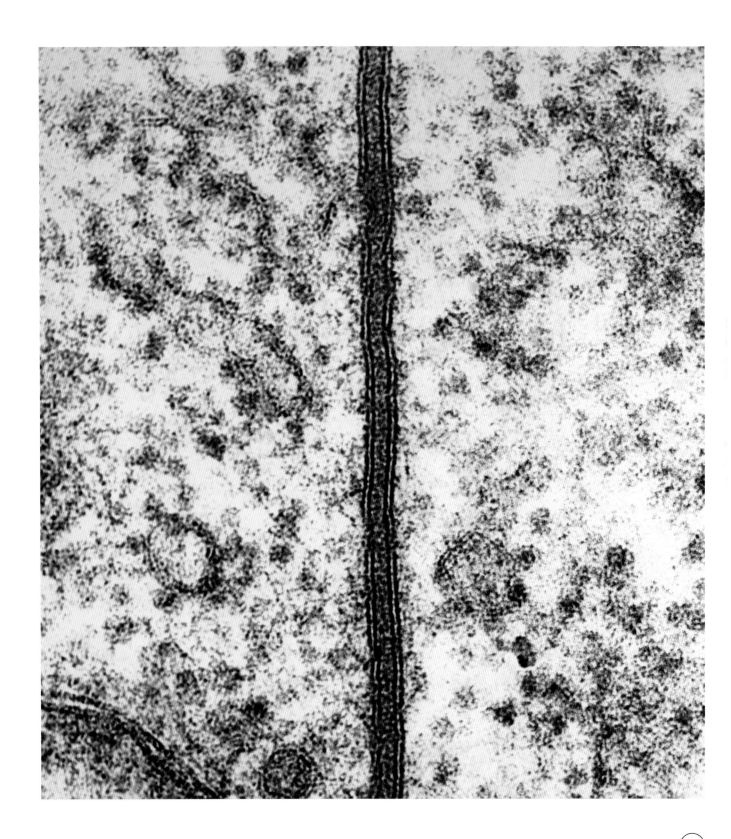

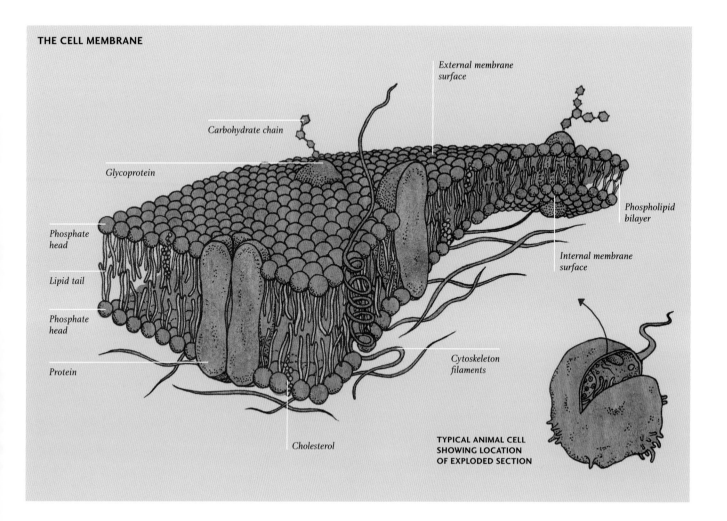

THE CELL MEMBRANE

TYPICAL ANIMAL CELL SHOWING LOCATION OF EXPLODED SECTION

on oxygen are known as anaerobic). Carbon dioxide is a waste product of cellular respiration and must leave the cell. Fortunately, these two small molecules, O_2 and CO_2, are fat-soluble; they pass through the fatty membrane relatively easily in a process called diffusion. They pass in both directions through the membrane at all times, but there will be more crossings from the side that has more molecules—so that, on balance, both oxygen and carbon dioxide pass from a higher to a lower concentration.

Molecular guest list

Most cells also need glucose, the fuel for cellular respiration. Glucose cannot diffuse through the phospholipid bilayer as oxygen molecules do. There are, however, transfer proteins that only let through glucose molecules, just as the aquaporin proteins allow through only water. There are similar transfer proteins that let other molecules through, each one specific to a particular compound.

Also embedded in the fluid mosaic of the cell membrane are receptor proteins. Each of these proteins allows a specific compound or ion—a ligand—to attach to it. Most ligands are proteins that have been released by other cells, such as hormones and neurotransmitters, and which, therefore, act as a signaling mechanism. There are, for example, special cells in the pancreas that release the hormone insulin when blood sugar is high.

When ligands "dock" on the part of the receptor sticking out of the membrane, the part of the receptor molecule inside the cell can change shape, start an enzyme reaction, or release compounds. These cause the cell to change its behavior. Human muscle cells and fat cells have insulin receptors in their membranes; when insulin docks into the receptors, the receptors directly affect the glucose transfer proteins. That encourages the transfer proteins to let more glucose into these cells for storage as glycogen or fat. The effect of all this is to reduce blood sugar level and to stock up on energy for times when food is scarce. It is a remarkably robust system, although it doesn't work in people with diabetes. In type-1 diabetes, the cells in the pancreas fail to produce enough insulin; in type-2 diabetes, the receptors stop responding adequately to the presence of insulin.

Like the bubbles in a glass of champagne

There is another way that substances can get in and out of a (eukaryotic) cell, a method perhaps even more remarkable than protein gatekeepers. Molecules that the cell needs can be "swallowed." When the desired molecules are next to the membrane, an indentation forms in the membrane; the indentation grows deeper and eventually part of the membrane completely surrounds the molecule. That portion of membrane snips off, forming a tiny delivery bubble, which is then subsumed down into the cell. Imagine a video showing a bubble rising to the surface of a glass of champagne and bursting … played in reverse. Once inside, the bubble melds into internal membranes, and the precious compound is released and made available.

The opposite also happens. Now play that champagne video forward to show bubbles bursting at the surface. Certain compounds manufactured inside the cell, mostly proteins, cannot pass directly through the membrane. Instead, they are carried up to the membrane inside a fatty bubble; this pops open at the surface, releasing the contents outside the cell. This is how insulin is released from those pancreatic cells; the compounds that build cell walls are also released this way. The fatty delivery bubbles originating from inside the cell are called "secretory vesicles" (there are other types of vesicle, see page 47). The precious protein cargo they carry is manufactured deep inside the cell, according to instructions held in the DNA in the nucleus. The process by which those instructions are copied and carried out is another incredible series of molecular events.

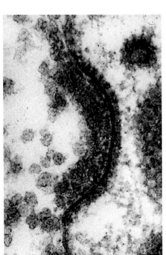
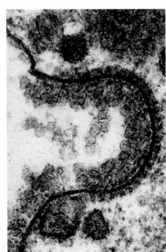
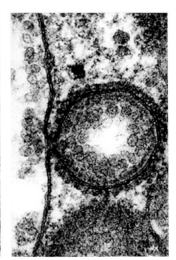
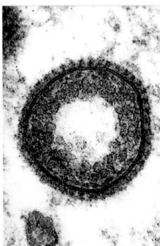

Below *Series of transmission electron micrographs that show the process of endocytosis—when a cell absorbs substances across its cell membrane. Receptor proteins in the membrane change shape when a target compound is present, forming an indentation and eventually a vesicle that surrounds the compound. The vesicle carries the compound into the cytoplasm.*

Protein factory

Proteins have many vital roles in living things. Some proteins are enzymes, which enable essential chemical reactions; some are structural materials inside or outside the cell; some are used as signaling molecules; and some as receptors. Then there are carrier proteins, such as the oxygen-carrying hemoglobin. All proteins are manufactured inside cells by incredible molecular machines called ribosomes, which follow instructions held in DNA.

Words of life

Proteins are made up of smaller molecules—molecules of compounds called amino acids. A protein molecule is not really a polymer, however, because the amino acids are not connected in a repeating pattern. Instead, they attach together one after another, like letters in a long word. The hormone insulin (in humans) is a protein consisting of a sequence of 51 amino acids. That's relatively short; the average length of a protein manufactured in human cells is 470 amino acids. The sequence is dictated by the protein-building instructions in DNA. Just as there is a fixed pool of letters in the alphabet from which to make words, so there are only 21 amino acids available to build proteins. That's more than enough, however, to build tremendous variety into the world of proteins. The number of possible combinations in a string of just ten amino acids is in the tens of trillions; most proteins are hundreds or thousands of amino acids long. Protein biochemists have identified around one million different proteins so far.

Unlike a word printed on a page a long chain of amino acids is rarely straight; most protein molecules bend and fold back on themselves. The shape of a protein molecule is often very important to its function. Signaling proteins, such as hormones, for example, fit into receptor proteins as keys do into

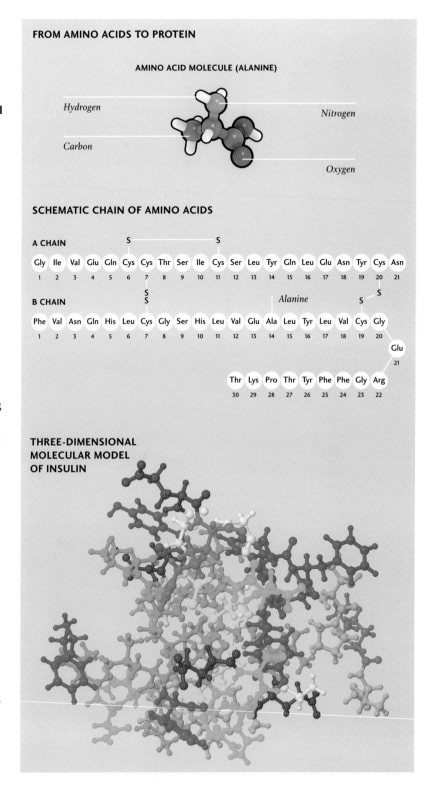

FROM AMINO ACIDS TO PROTEIN

AMINO ACID MOLECULE (ALANINE)

Hydrogen
Carbon
Nitrogen
Oxygen

SCHEMATIC CHAIN OF AMINO ACIDS

THREE-DIMENSIONAL MOLECULAR MODEL OF INSULIN

INSIDE LIVING CELLS

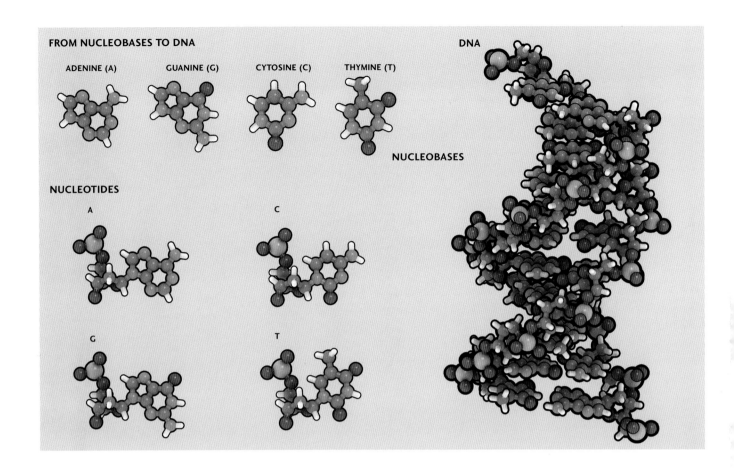

Above The DNA double helix is made of nucleotides, each of which is a nucleobase plus a deoxyribose sugar and a phosphate. The sugar and phosphate groups form the twisting "backbones" of the DNA helix, while the nucleobases form the "rungs" that join the backbones.

Left Amino acids are made up of carbon, oxygen, nitrogen, and hydrogen atoms. A sequence or chain of amino acids join together to make a protein. In the model, each amino acid is shown in a different color. The protein is folded and, here, cross-linked by sulfur atoms (S), and the resulting shape is important.

locks. Many proteins change shape, or undergo conformational changes, when their circumstances change—for example, when certain other molecules are nearby or when the acidity or temperature changes. This is an important part of how cells respond to their environment.

Protein recipes

A gene is a particular section of DNA that represents the "recipe" for a particular protein. Genes exist one after another along the length of DNA, and each gene is "written" in code, as a sequence of nucleobases. There are four nucleobases in the DNA molecule: adenine (A), guanine (G), thymine (T), and cytosine (C). Each nucleobase is attached to part of the "backbone" of the DNA molecule, a sugar-phosphate group. The resulting structure is called a nucleotide. The four nucleotides are also referred to as A, G, T, and C, according to which nucleobase is present. Because these pieces are the building blocks of DNA, biologists talk about nucleotides more than they do nucleobases.

There is a direct connection between the A, G, C, and T nucleotides of DNA and the 21 amino acids of proteins. The nucleotides exist in sets of three—called triplet codons. With three slots and four possible nucleotides per slot, altogether there are 64 combinations (4 x 4 x 4)—more than enough to code for the 21 amino acids and to include administrative tasks, such as noting the start and end of a gene. Codon ATG means the start of a gene, while TAT represents the amino acid tyrosine.

The DNA molecule is a double helix, like a twisted rope ladder. Two strands twist around each other, zipped together by bonds between the nucleobases along the length of each strand. The two strands are complementary: A always bonds with T, and C always bonds with G. So a section of DNA that reads ATCGTA on one strand will be joined to the sequence TAGCAT on the complementary strand. Only one strand holds the actual code, but the other strand is not redundant; it plays a very important role, as we will see.

The length of a section of DNA is normally given as a number of base pairs—each base pair being a single rung on the twisted rope ladder. That same measurement could just as well be bases or nucleotides. Genes are generally a few thousand base pairs (kbp) long. The gene for a protein called dystrophin, found in muscle, is the longest known; it is 2.4 million base pairs long. The entire length of DNA of an organism, copied in each cell (well, most of them), is that organism's genome. Generally—although not always—the more complex an organism, the more base pairs will make up its genome. Bacterial genomes can range from around 150 thousand to about 12 million base pairs in length. In eukaryotes the number is much greater; for example, the human genome is about three billion base pairs long. A good deal of a genome's DNA has no (or no known) function.

Loops, beads, and shoelaces

A single length of DNA, end to end, is called a chromosome, and it contains many genes along its length. The prokaryotic cells of Bacteria and Archaea have a single chromosome, arranged as a closed loop that floats free in the cytoplasm. Most eukaryotic cells have between several and many chromosomes, which are held inside the nucleus (see box). For most organisms, the genome

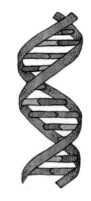

Above *Schematic diagram of a section of DNA. The green twisting section represents the sugar-phosphate backbone, and the rungs are the nucleobases. A (purple) always bonds with T (orange), and C (yellow) with G (pink).*

THE NUCLEUS

The function of the nucleus (found only in eukaryotic cells) is simply to keep the huge expanse of DNA in one place. The nucleus has a phospholipid bilayer membrane of its own, just like the cell's main membrane. And, of course, it lets things in and out, just as the cell's membrane does. There are two thousand or so pores in the nuclear membrane, each one a collection of proteins with a centered hole in it, embedded in the membrane. Various compounds pass in and out of the nucleus through these membranes, including the ingredients for making RNA, DNA, and proteins.

Around certain sections of DNA inside the nucleus is a dense crowding of molecules that constantly produce new ribosomes, the molecular machines that actually build the proteins. This densely packed region is called the nucleolus.

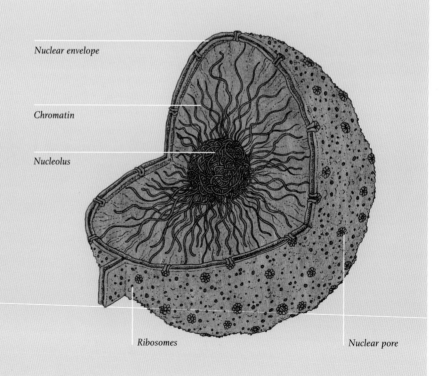

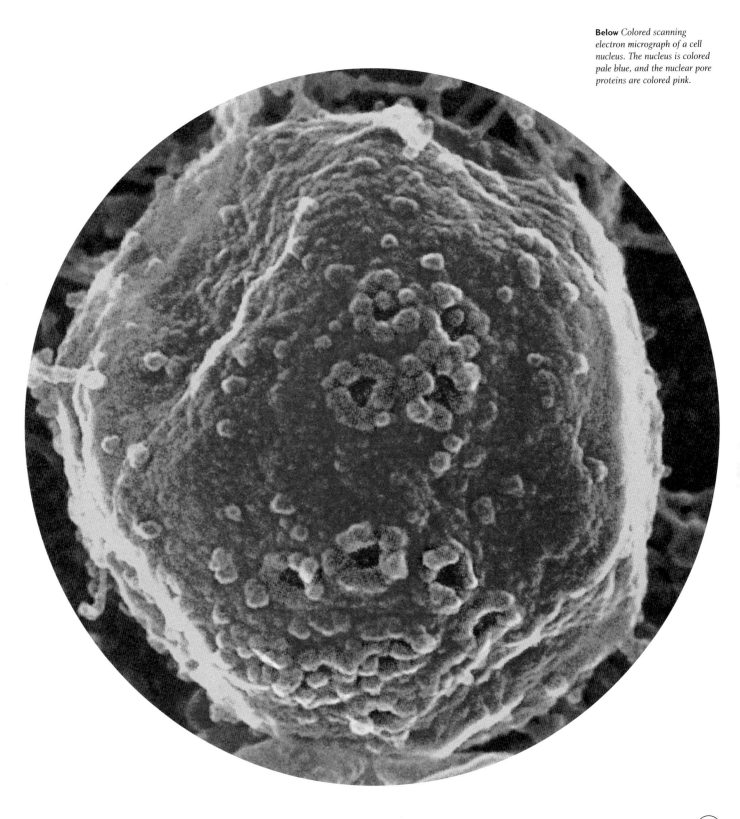

Below *Colored scanning electron micrograph of a cell nucleus. The nucleus is colored pale blue, and the nuclear pore proteins are colored pink.*

is a huge length of DNA—in humans this amounts to a total of about 6 feet (2 meters)—in each cell. In order to fit all that in the nucleus, the DNA has to be very tightly packed.

In eukaryotic cells (and in archaea), the DNA is wrapped every so often along its length around blobs of a protein called histones; the result looks like beads (histones) on a string (DNA). The beaded string is further wrapped around itself, so that it becomes a much fatter string—although still just 30 nanometers (0.00003 millimeters) in diameter. So each chromosome exists as a 30-nanometer fiber, chaotically jumbled with the other chromosomes, like so many shoelaces in a bag. Order is imposed on chromosomes during cell division, however, when the chromosomes organize themselves into discrete packages, ready for duplication. This is a process that is developed in chapter three.

When a protein is to be made, enzymes unpack and unwind the section of the DNA containing the relevant gene, so that the code becomes exposed. It marks the beginning of an amazing process of molecular recipe-copying, protein assembly, modification, and delivery.

Right *Transmission electron micrograph of chromatin, which is comprised of DNA wrapped around histone proteins. Here, in the cell nucleus, most of the chromatin is coiled around itself, forming 30-nanometer (0.00003-mm) fibers.*

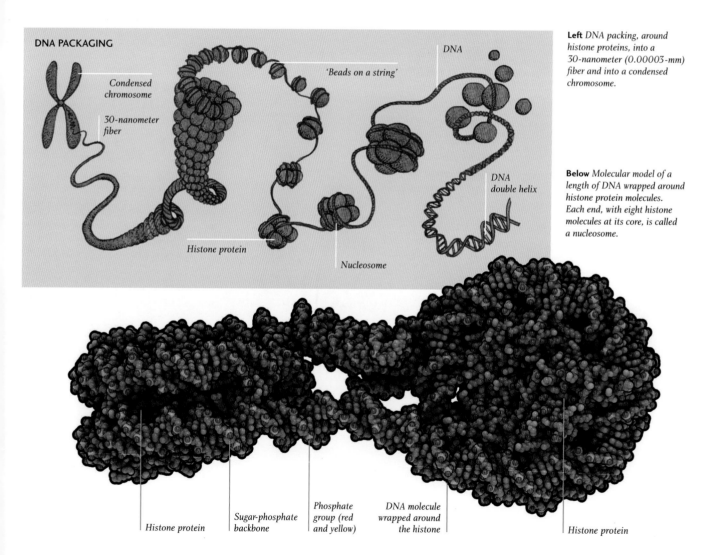

Left *DNA packing, around histone proteins, into a 30-nanometer (0.00003-mm) fiber and into a condensed chromosome.*

Below *Molecular model of a length of DNA wrapped around histone protein molecules. Each end, with eight histone molecules at its core, is called a nucleosome.*

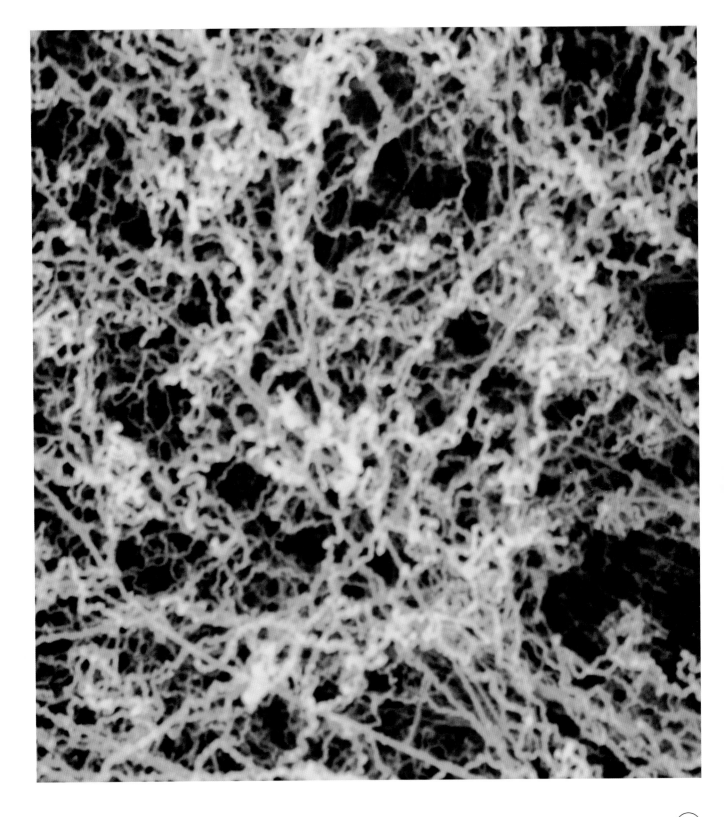

Writing out the recipe: transcription

The DNA holds the key for protein building, but it doesn't do the building, nor is the building done inside the nucleus. The first two stages in the process of protein building are copying out the recipe and taking it outside the nucleus. The copy is written as a length of RNA (ribonucleic acid), a compound very similar to DNA (deoxyribonucleic acid) but single-stranded. The RNA is copied from the complementary strand of DNA, not the coding strand. In this way, it is a copy of the coding strand—and that is why the second, complementary strand of DNA was so vital.

The "scribe," which does the copying, is an enzyme—RNA polymerase. Before its job of transcription begins, a cluster of other molecules attaches to the DNA and prepares it for copying. The RNA polymerase then unzips the DNA double helix and makes a complementary copy of the complementary DNA strand from a multitude of nucleotides floating in the nucleus. The RNA is not exactly the same as the coding DNA of which it is a copy: all the T nucleotides have become U (the nucleobase called uracil), but the code still reads just as it did.

The RNA polymerase zips the DNA back up behind itself. As it moves swiftly along the DNA, a growing ribbon of RNA emerges from the side. When it reaches the end of the gene, the RNA polymerase molecule detaches, ready to read another gene, and the freshly made ribbon of RNA is let loose, ready for the next stage. This precious piece of genetic code is called messenger RNA (or mRNA), for it will carry the copied recipe to the manufacturing stage. Gatekeeper proteins embedded in the nucleus's membrane encourage the mRNA out of the nucleus.

Do remember that this fundamental and intricate process is self-organized, not under any kind of central control. It is just a set of complex chemical reactions.

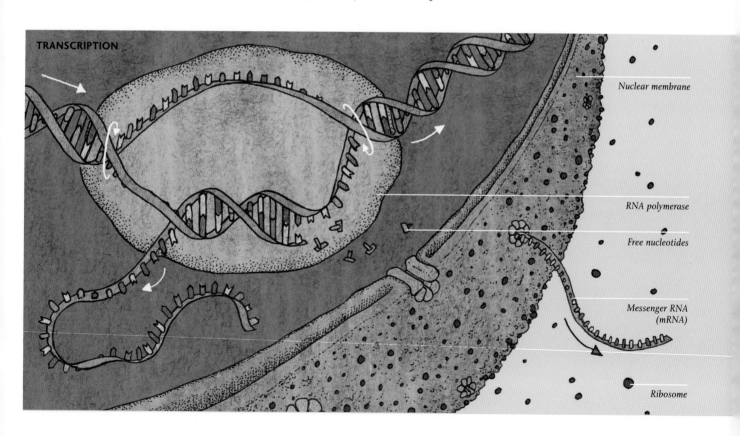

INSIDE LIVING CELLS

Reading the recipe: translation

So, a piece of messenger-RNA has just left the nucleus. This copy of the DNA's recipe for a protein must now be translated into amino acid language. The protein has to be built up as a specific sequence of amino acids. Welcome to the remarkable process of protein translation.

A self-assembling molecular machine called a ribosome clamps onto the new ribbon of mRNA. The mRNA code passes codon-by-codon (three nucleotides at a time) through the heart of the ribosome—and smaller molecules now dock there, bringing one amino acid each. These new molecules are small lengths of transfer RNA (tRNA), so-called because they transfer the amino acids from the cell at large into the ribosome to assemble the protein. Each tRNA is just three nucleotides long, so that it corresponds to a particular codon. The amino acid it carries at its head corresponds to the codon it represents at its foot. Again, it is worth remembering that this is just a chemical reaction.

The three-toed foot of the tRNA molecule docks with the relevant codon on the mRNA at the heart of the ribosome … and then the next one sidles up next to it and does the same. The two amino acids join, and the ribosome moves to the next codon. The tRNA, now emptied of its load, is released and will soon be endowed with another identical amino acid, ready for its next transfer.

And so, just as a ribbon of mRNA emerged from the RNA polymerase as it moved along, the DNA inside the nucleus, now a growing chain of amino acids—a protein molecule—emerges from the ribosome as it makes its way along the mRNA. This amazing process is similar in prokaryotic cells, but the ribosomes and the step-by-step details are slightly different than those in eukaryotic cells.

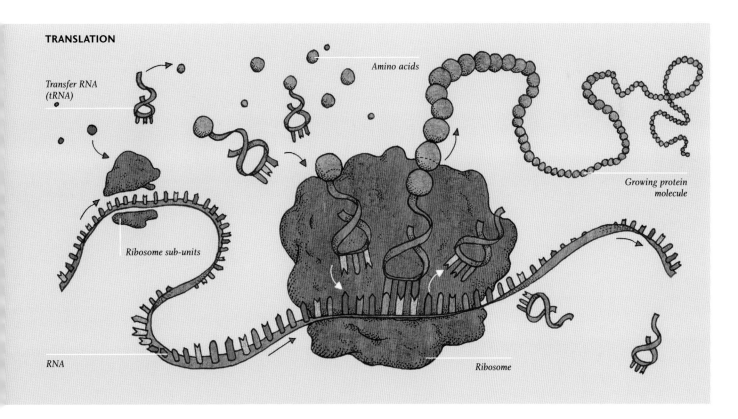

TRANSLATION

Transfer RNA (tRNA)

Amino acids

Growing protein molecule

Ribosome sub-units

RNA

Ribosome

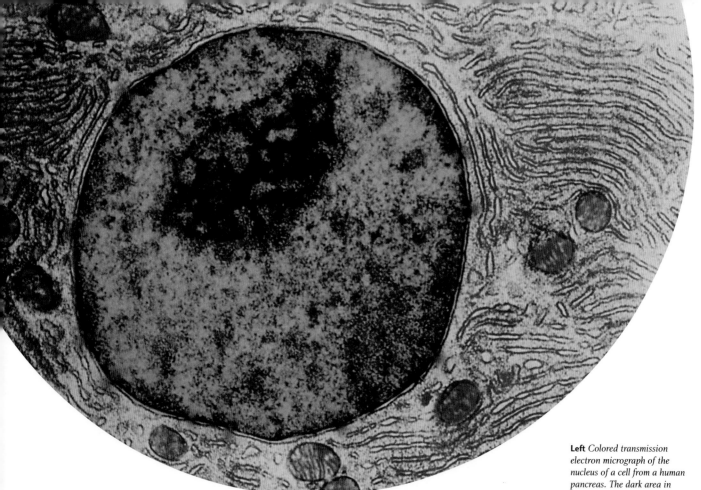

Left *Colored transmission electron micrograph of the nucleus of a cell from a human pancreas. The dark area in the nucleus is the nucleolus, a region active in making mRNA. Nuclear pores, highlighted in red, let mRNA molecules out. The endoplasmic reticulum, studded with ribosomes, is colored dark blue. (The pink-colored structures are mitochondria.)*

Taking the rough with the smooth

The membrane of the nucleus is a phospholipid bilayer, just like the cell membrane. Attached to the nucleus's envelope, and extending from it, is yet another bilayer membrane structure, called the rough endoplasmic reticulum (RER). It is rough because it has ribosomes embedded in it, and it is endoplasmic because it is in the inside (endo-) of the living mixture (plasm) of the cell. And reticulum is just a fancy word for net. The RER looks a little like a net, although it is actually more like a maze that might be included in a puzzle book.

Ribosomes become embedded in the RER if a particular code is present in the mRNA protein recipe. Otherwise they do their protein manufacturing floating free in the cytoplasm.

Proteins manufactured on the surface of the RER may have sugar molecules added, or may be better folded into the correct shape when made here. Most are destined for release from the cell inside a secretory vesicle, one of those champagne bubbles that will burst out of the cell membrane (see the box opposite).

Farther out but also connected to the nuclear envelope is another folded membranous structure: the smooth endoplasmic reticulum (SER). As its description suggests, this is not studded with ribosomes—so, unlike the RER, it has no role in protein building. Instead, lipids, phospholipids, and cholesterol and other steroids are made here. So cells that make steroid hormones, such as testosterone and estrogen, have more SER than

VESICLES

The champagne bubble vesicles that deliver secretory proteins to the cell membrane are not the only kind of vesicle present inside cells. Plant cells have huge, membrane-bound spaces called vacuoles, which are really just large vesicles. The main function of a vacuole is to store useful—and potentially toxic—compounds. A vacuole also stores water; it can swell to occupy 95 percent of the cell's volume. It stores pigments and also poisons that can protect the plant if the cell is attacked. For example, vacuoles inside the cells of a garlic bulb hold allicin, the sulfurous compound that gives garlic its flavor. A plant cell's waste products are also processed inside the vacuole.

Plant cells have vacuoles, but the equivalent vesicles inside animal cells are called lysosomes. Inside the membrane of a lysosome, obsolete and toxic molecules are broken down into their constituent parts, which are then released for recycling if they are useful. It is in lysosomes that proteins are broken down into their constituent amino acids, for example, and released ready for ribosomes to use them to make new proteins.

most other cells. It is also involved in breaking down toxins, so the detoxifying cells in the liver also have more SER than most other cells.

Delivering the goods

Proteins that were manufactured by ribosomes embedded in the RER are now dispatched to another membranous structure: the Golgi apparatus. They travel in style, enclosed in vesicles. The Golgi apparatus is similar to the ER; it is made up of stacked layers of membranes. As a protein passes from layer to layer on its way up toward the cell's outer membrane, it is altered in any of a number of ways. Its sugar molecules might be changed; its shape may change again; in some cases, part of the protein may be cut off, transforming it from an inactive to an active state. This last action, for example, is what happens to insulin as it passes through the Golgi apparatus.

Once through the layers of the Golgi apparatus, the secretory protein arrives at the cell membrane, ready to be released in a graceful bursting of the bubble. Often, however, proteins will be stored in the Golgi apparatus so that they can be released quickly. This is particularly true of neurotransmitters in neurone cells. When the cell receives a particular signal, the neurotransmitter proteins are released from vesicles that rapidly burst at the cell's surface, ready to help make a new memory, play a part in perception, or cause a muscle to contract.

Left *Colored transmission electron micrograph of yeast cells. The cell wall is colored green, and in the center is a kind of vesicle called a vacuole (pale blue). All fungal cells have vacuoles, whose main purpose is to hold water with either waste products or important compounds in solution.*

Frames and motors

The vesicles that act as delivery vehicles inside eukaryotic cells don't just float around randomly. Instead, proteins walk them along rigid tubular tracks inside the cell. These tracks are part of an extensive network of protein fibers known as the cytoskeleton.

The power of three

The cytoskeleton is found only in eukaryotic cells. It has several crucial roles beyond providing delivery tracks. It makes cells move; it supports organelles; it drags chromosomes around in the dance of cell division (the music for which begins in chapter three); and in cells with no cell wall, it determines the overall shape of the cell. The cytoskeleton is made of three different types of protein fiber, which can be considered as small, medium, and large. The names (or at least the first two) help: microfilaments, intermediate filaments, and microtubules. The smallest and largest—the microfilaments and microtubules respectively—are dynamic structures, constantly assembling and disassembling from smaller pieces called subunits. The intermediate filaments are less dynamic.

Microfilaments are extremely slender protein threads with a diameter of about 6 nanometers (0.000006 millimeters)— about one-thousandth the diameter of a typical human hair. Microfilaments have tensile strength; they are strong under tension, like the guy ropes of a tent. By pulling, they can change the shape of a cell—for example, extend an amoeba's "foot." In the cells that line our intestines, the microfilaments extend into tiny bristles called microvilli, which dramatically increase the surface area of the cells, enabling them to absorb more

Below left *Light micrograph of a fibroblast cell, showing the actin filaments of the cytoskeleton, stained purple.*

Below right *Light micrograph of microtubule filaments, made of the protein tubulin, which has taken up a red fluorescent stain (rhodamine).*

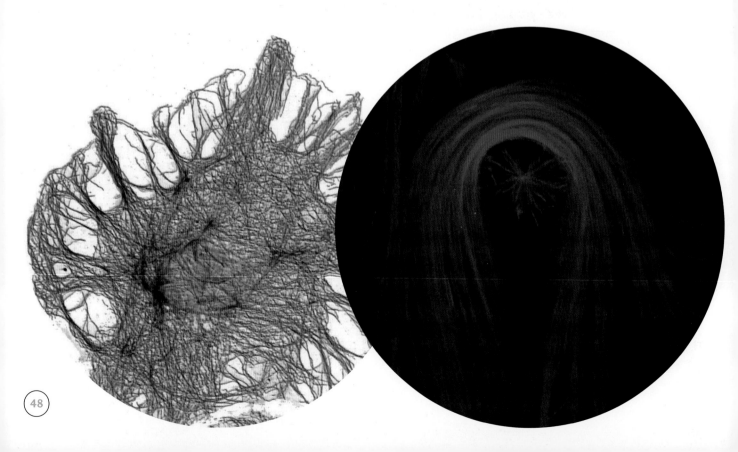

Right *Individual molecules (monomers) connect together to form a repeating pattern (polymer). The G-actin molecules (green) are bound to ATP (yellow), and the resulting long, filamentous polymer resists tension.*

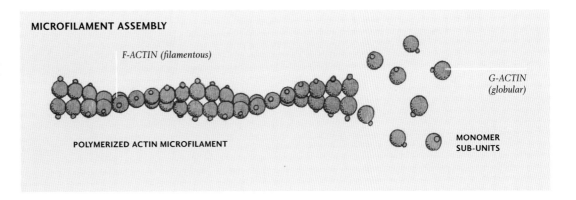

MICROFILAMENT ASSEMBLY

F-ACTIN (filamentous)

G-ACTIN (globular)

POLYMERIZED ACTIN MICROFILAMENT

MONOMER SUB-UNITS

Below left *Fluorescent light micrograph of two fibroblast cells. Nuclei are purple, microtubules yellow, and actin microfilaments pale blue.*

Below right *Colored transmission electron micrograph of a desmosome (dark green) attached to an intermediate filament called a tonofibril, which anchors the desmosome to the cytoskeleton.*

nutrients. Microfilaments also play a central role in the contraction of muscle cells, which will feature in chapter seven.

The intermediate filaments are medium in diameter and also in rigidity. They are like thick twisted ropes rather than thin fibers or rigid tubes. There are several kinds, each made of different proteins. Overall, they are responsible for maintaining a cell's shape, including keeping the other two types of fiber in place.

One kind of intermediate filament, made of the protein lamin, forms a grid that gives shape to the membrane around the cell nucleus. The toughness of skin and hair is due to tangled, cross-linked intermediate fibers made of keratin. Yet another kind of intermediate fiber forms structures on the cell membrane that connect to similar structures in neighboring cells. These structures are called desmosomes, and in animals they hold cells together to form tissues, such as muscle and mucous membranes.

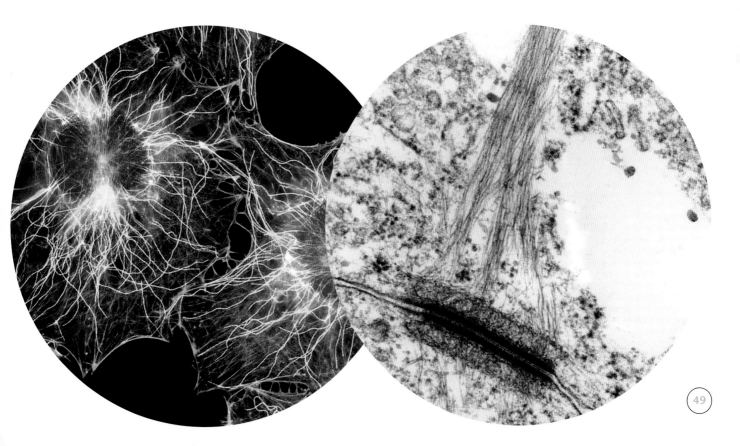

Delivering the goods

It is the largest fibers, the microtubules, that form the remarkable network of tracks enabling vesicles to be delivered around the cell. Microtubules are rigid, hollow rods made of a protein called tubulin, which can disassemble as quickly as they can assemble. Small tubulin subunit molecules are everywhere, dissolved in the cytoplasm, ready to be used in a kind of cellular construction kit. The motor proteins that walk along the tracks are called kinesins and dyneins. Often the track is made just in front of the approaching kinesin, like laying down railroad tracks in front of a moving train.

The kinesins always walk towards the cell membrane, the dyneins away from it. The gait of these mobile molecules is a little awkward, but they race along, taking up to one hundred steps every second. One hundred motor protein steps move the proteins about 800 nanometers—which works out as a few millimeters per hour. The cargo, securely attached to the head of these incredible walking proteins, is not restricted to vesicles: it can be RNA, proteins, or even entire organelles. To move a particularly large organelle, two or more "walkers" team up—especially if there is some resistance to the organelle's movement.

Whips and eyelashes

Microtubules extend into long, thin, hairlike extensions called cilia (the name is Latin for "eyelashes"). Many cilia carry the sensing part of sensory cells in the nervous system; this is true of the cells that give us our sense of smell (olfactory neurons) and the light-sensitive cells in the retinas of our eyes, which will feature in the final chapter. Kinesins and dyneins carry nutrients and other essentials to and fro inside each cilium, along its length. They play an even more dynamic role in some cilia, making them move back and forth. In the lining of our lungs and windpipe, for example, cilia are flapping back and forth between 10 and 20 times every second, pushing mucus-soaked dust particles and bacteria up to the top of

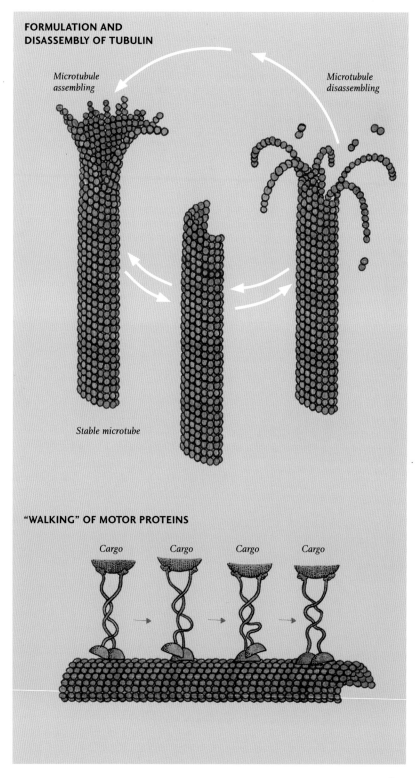

FORMULATION AND DISASSEMBLY OF TUBULIN

Microtubule assembling

Microtubule disassembling

Stable microtube

"WALKING" OF MOTOR PROTEINS

Cargo

INSIDE LIVING CELLS

Left Microtubules (top) are built from units each consisting of two proteins, alpha- and beta-tubulin. They are dynamic structures, with even apparently stable tubes constantly being assembled and disassembled. Motor proteins (bottom) move along the surface of microtubules in a walking action.

Below left Inside flagella and cilia microtubules are arranged in neat geometric bundles.

Below right Scanning electron micrograph showing the root of the two flagella of Chlamydomonas reinhardtii, a single-celled green alga (a eukaryote).

the windpipe, where they can be swallowed. And the human egg is teased along the Fallopian tubes by flapping cilia. Inside these motile cilia, the kinesin and dynein grab hold of a centered column and walk up and down, so that the whole column of tubulin bends.

While only eukaryotic cells have a cytoskeleton and cilia, there is a related structure that is found in some bacterial and archaeal cells as well as eukaryotic ones: the flagellum, named after the Latin for "whip." Flagella are much longer and fewer in number than cilia, but they work in a similar way. Eukaryotic flagella have a framework of tubulin and work just as motile cilia do. There is a large flagellum located on the back of each sperm cell, pushing it along like a motorboat. The framework inside a bacterial flagellum is made of tubes of a similar protein, called flagellin. Rather than waving to and fro, this flagellum is made to rotate, at a rate of several hundred revolutions per minute, by a molecular motor at its root. These motors are very effective: a waterborne bacterium can move up to 50 times its own length every second—the equivalent of a person swimming at about 100 mph (160 km/h).

The energy required to build the cytoskeleton, to power the motor proteins that walk along the microtubules, and to drive flagella comes from cellular respiration. The energy is delivered in little packages of a compound called adenosine triphosphate (ATP). The "walking" proteins kinesin and dynein require one ATP for each step, up to 100 per second.

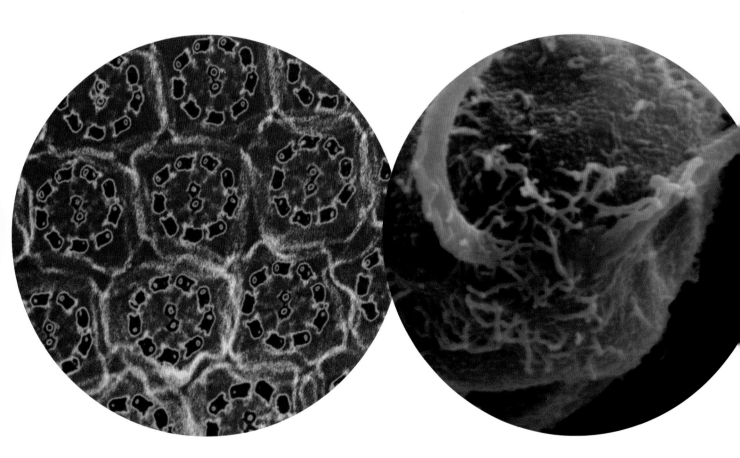

INSIDE LIVING CELLS

Feeding the fire

All cells obtain the energy they need to survive, grow, and reproduce from energy-rich chemical compounds, such as glucose. Some cells are able to manufacture their own glucose, using the energy of sunlight.

Invest to spend
The two main processes by which energy is released from cellular fuels, such as glucose, are respiration and fermentation. There are several different variations on these two themes, but in every case the energy released by processing cellular fuels ends up being stored in cells' universal currency—molecules of adenosine triphosphate, ATP.

The ATP molecules are built up from lower-energy adenosine diphosphate (ADP) by adding a phosphate group, an operation that requires energy. It may help to think of ADP as an empty wallet and ATP a full one, with the extra phosphate as the cash that fills that second wallet. ATP molecules at large throughout the cell spend their energy powering a multitude of different processes. In the most common and most efficient form of respiration, called aerobic respiration, each glucose molecule ultimately yields about 36 ATP molecules.

Burning up
Both respiration and fermentation involve oxidation. A classic, readily understood example of oxidation is combustion, or burning. When wood burns, for instance, oxygen from the air combines with carbon and hydrogen atoms from the wood, producing carbon dioxide, CO_2, and dihydrogen oxide (better known as water, H_2O), and releasing energy in the form of heat. However, despite the name, oxidation doesn't necessarily involve oxygen at all. The process is actually about the transfer of electrons: atoms or molecules become oxidized when they lose electrons.

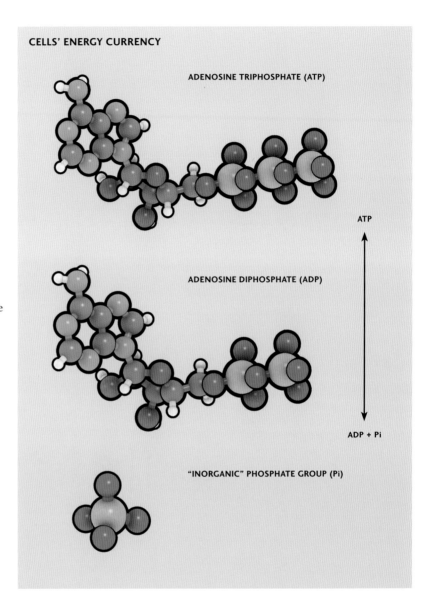

CELLS' ENERGY CURRENCY

ADENOSINE TRIPHOSPHATE (ATP)

ADENOSINE DIPHOSPHATE (ADP)

"INORGANIC" PHOSPHATE GROUP (Pi)

ATP ↕ ADP + Pi

Right *Transmission electron micrograph of a mitochondrion showing the double membrane. Visible to the left of the membrane is another membranous organelle—the rough endoplasmic reticulum.*

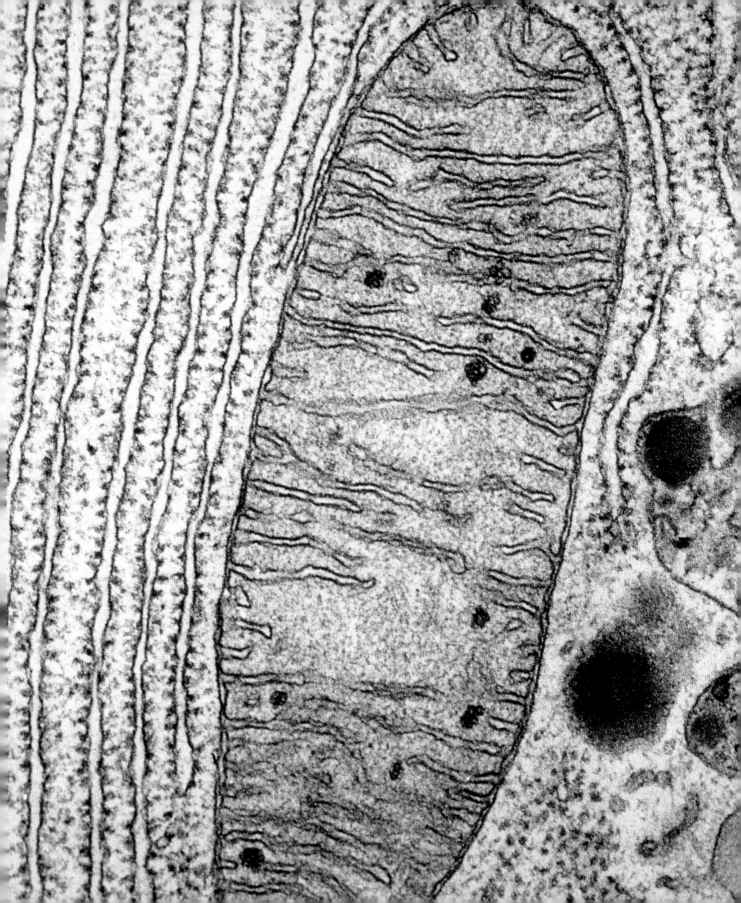

The opposite of oxidation is reduction, which happens whenever elements and compounds gain electrons. Inside iron ore, for example, each iron atom is missing two or three electrons and is, therefore, a positively charged iron ion. Smelting the ore involves reducing the iron ions by reacting the ore with electron-donating elements, such as carbon. Once pure, the iron slowly reacts with air and water—both act as oxidizing agents, stealing electrons from the iron atoms. The result is that the pure iron becomes oxidized as rust (a mixture of iron oxides and iron hydroxides). Reducing an element or compound requires an input of energy, and oxidation releases that energy. The carbon atoms in fats and proteins and in carbohydrates, such as glucose, are all in a reduced state, having been made that way by metabolic reactions inside cells.

Something in the air

Aerobic respiration is a three-step process. It begins when glucose is split in two; this step (glycolysis) actually requires energy. The rewards soon start coming in, however: in the next step, a complex series of reactions taps much of the available energy. This step, called the Krebs cycle or citric acid cycle, produces an array of intermediate compounds. Ultimately it provides several molecules of ATP and a lot of molecules of another high-energy compound called nicotinamide adenine dinucleotide (reduced form)—thankfully known by its acronym, NADH. Most processes in the cell recognize only ATP, so the last step is a kind of currency conversion.

The ATP produced in the final step is constructed by a very special enzyme, ATP synthase, which sits in a membrane. This step is called the electron transport chain. Electrons from NADH are carried back and forth across the membrane, each time "pumping" hydrogen ions, aka protons, from one side of the membrane to the other. The resulting pressure pushes protons back across the membrane, through the ATP synthase molecule, making one part of it literally rotate. The energy made available in this way unites ADP molecules with spare phosphates to make molecules of ATP. This remarkable process is called oxidative phosphorylation.

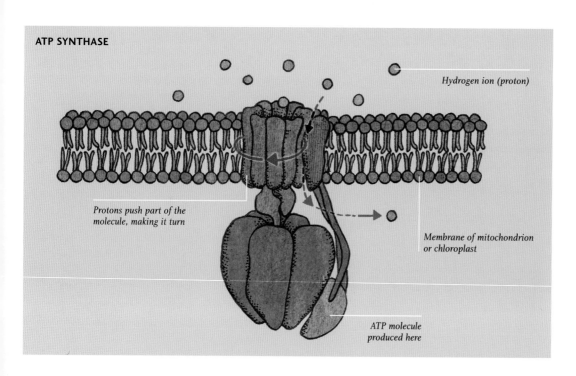

ATP SYNTHASE

Hydrogen ion (proton)

Protons push part of the molecule, making it turn

Membrane of mitochondrion or chloroplast

ATP molecule produced here

LIVING WITHOUT OXYGEN

Some cells get by without oxygen—either as their only form of respiration or as a way of carrying on respiration when no oxygen is available. Anaerobic respiration is very similar to aerobic respiration but uses alternative oxidizing agents, such as sulfur or nitrates. Nearly all anaerobes are bacteria or archaea.

The other main alternative is fermentation. This involves the breakdown of compounds produced by what is normally step one of cellular respiration. The products of fermentation are normally ethanol (alcohol) and carbon dioxide gas—putting the fermentation process at the heart of brewing and baking. Sometimes lactic acid is produced; an athlete's overworked muscle cells produce lactic acid when oxygen is just not arriving fast enough, for example, and the buildup of this acid causes pain in hardworking muscles. Lactic acid produced by fermenting bacteria curdles milk, a vital step in making yogurt.

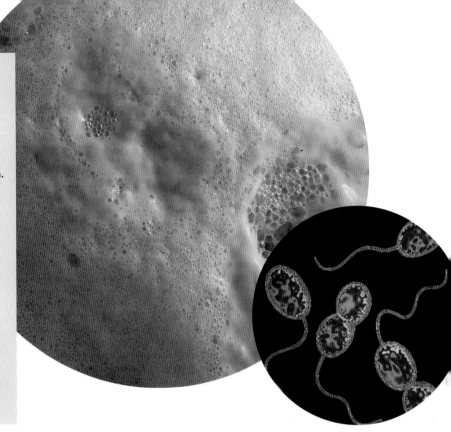

Above left *Bubbles of carbon dioxide gas produced by yeast accessing the energy in sugar via fermentation.*

Above right *Colored transmission electron micrograph of the sulfur-eating bacteria* Thiocystis. *These bacteria live in sulfur-rich lakes with little or no oxygen.*

The reaction needs a final step to mop up the now low-energy electrons that carried protons across the membrane. Something that mops up electrons is an oxidizing agent; in aerobic respiration, that agent is oxygen. It steals the electrons, which unite with used protons (hydrogen ions) that have come back through the membrane, via the ATP synthase molecule, to make molecules of water. There is a range of options available for cells that live in, or find themselves in, an environment with little or no oxygen (see box).

In the membrane

When aerobic respiration takes place inside prokaryotic cells, such as aerobic bacteria, ATP synthase sits in the actual cell membrane. This is part of the reason why bacteria have such a protective cell wall, or even a second membrane outside the main one. Eukaryotic cells, however, contain specialized, membrane-bound organelles called mitochondria. The second and third stages of cellular respiration both occur inside these power stations. Mitochondria have an inner and an outer membrane—and it is the inner membrane that is studded with ATP synthase molecules.

A mitochondrion is like a self-contained cell-within-a-cell. Its outer membrane acts in precisely the same way as the main cell membrane. Oxygen diffuses into the mitochondria, and the products of the first stage of cellular respiration (glycolysis) are taken in through selective membrane proteins. Similarly, carbon dioxide diffuses out, and the freshly produced ATP passes through membrane proteins. Mitochondria even contain their own DNA and their own ribosomes. These facts (and others) strongly suggest that mitochondria originated as free-living bacteria and that, millions of years ago, one of them entered into a beneficial relationship (symbiosis) inside (endo-) a host cell. This theory of endosymbiosis applies equally to another membrane-bound organelle found in plants and algae: the chloroplast.

Catching the sun
(Eukaryotic) plants and algae use energy in sunlight to manufacture glucose in the process called photosynthesis. They use the glucose as food—fuel for aerobic respiration—so they are self-feeders, or autotrophs. They also use glucose as the starting point for building energy-storage molecules, such as starch, and structural polymers, such as cellulose. Photosynthesis also takes place in some prokaryotic cells but, unlike plants and algae, they do not possess chloroplasts.

Inside, a chloroplast has its very own DNA and ribosomes, evidence of endosymbiosis, as it is in the case of mitochondria. Also like mitochondria, much of the action takes place on membranes that are studded with ATP synthase—and, once again, high-energy electrons pump protons across one side of the membrane to make ATP. But in a chloroplast the electrons come from water. Light energy captured by a pigment called chlorophyll splits water molecules, producing hydrogen ions (the protons) and high-energy electrons. The water's oxygen is released. The other ingredient of photosynthesis is carbon dioxide, which has been "breathed in" from the air. In a series of reactions called the Calvin cycle, which involves the electrons and hydrogen ions from the first step, the carbon dioxide is reduced to form glucose, the final product of photosynthesis. To summarize a complex series of events into a simple statement, photosynthesis is water and carbon dioxide in, glucose out. Of course, there are variations on that theme, too: just as there are some organisms that do not use oxygen in respiration, there are some autotrophs that do not use water in photosynthesis. The equivalent to anaerobic respiration is "anoxygenic photosynthesis."

All the remarkable happenings we have glimpsed in this chapter would be for nothing were it not for the things that we will glimpse in the next—for it is there we will look at how cells reproduce.

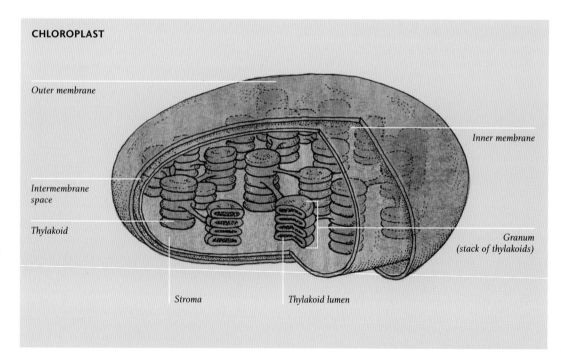

Right *Colored transmission electron micrograph of two chloroplasts in a plant cell. The grana (stacks of membranes, yellow) hold the chlorophyll molecules that absorb light energy. The black areas are starch granules, made from the sugars produced by photosynthesis.*

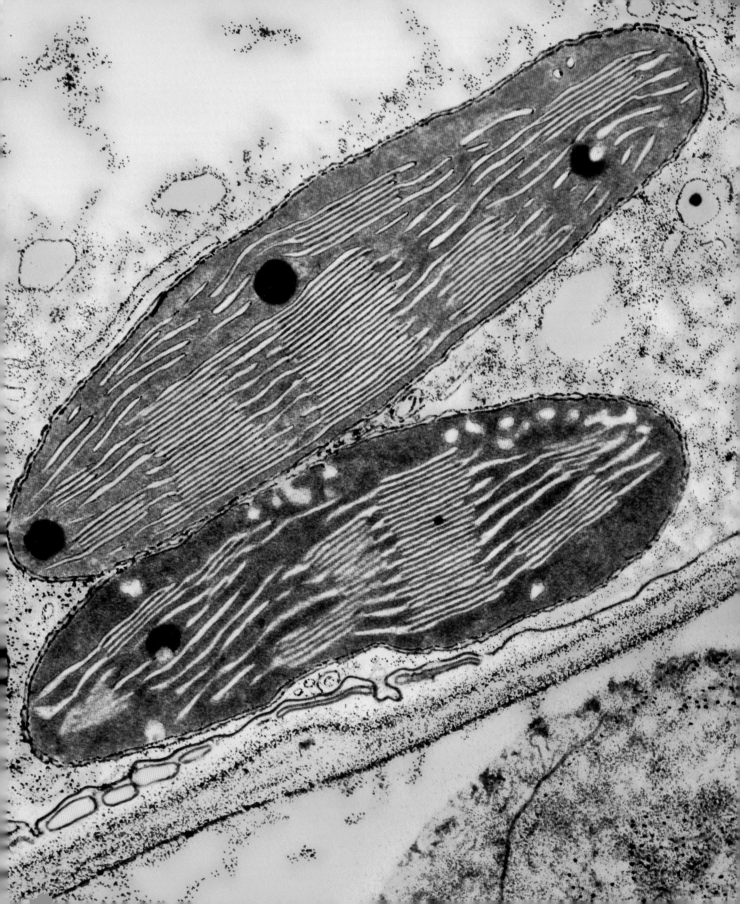

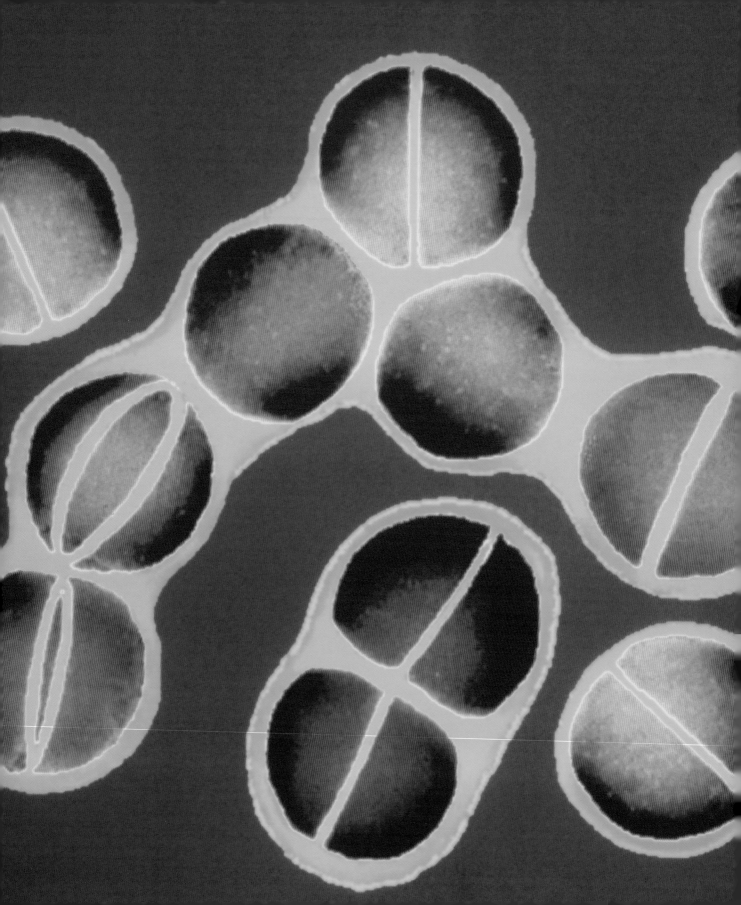

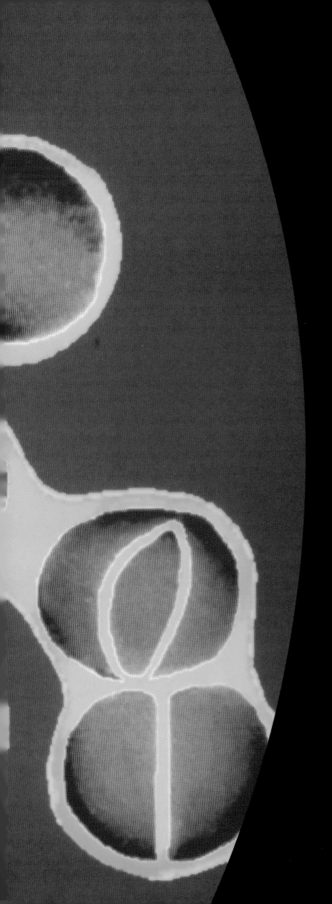

CHAPTER THREE
Cells Beget Cells

Here is a big thought. Every single single-celled living thing, and every single cell in every multicellular living thing, was created from an existing cell that split in two. The process of cell division is clearly important—and, by looking at the details, it becomes apparent that cell division explains far more than just how new cells are made.

Left *False color transmission electron micrograph of* Staphylococcus aureus *bacteria, most of which are in the act of dividing in two. Cell division is the way in which all single-celled organisms reproduce, and the way in which multicellular organisms grow.*

Re: Cycle

The incredible goings-on inside cells—the manufacture of proteins, the production of energy, the constant shuffling of motor proteins along microtubule tracks—would achieve nothing if cells did not have a way of reproducing by dividing in two. Eukaryotic cells divide in two via a chain of events called the cell cycle.

Grow, duplicate, split, repeat

Cell division is how single-celled organisms reproduce. Multicellular organisms grow and develop by repeated cell division, although some can also use cell division to reproduce (see box below). In order to divide successfully, a cell must ensure it contains enough vital material to make two cells, including making a copy of its genome (all of its DNA), and divide everything up equally. Once the cell has divided, the whole process begins again in each of the two "daughter" cells. In single-celled prokaryotic cells, cell division is simple and quick (see box on facing page). The process takes much longer in eukaryotic cells because their genomes are typically hundreds or thousands of times the size of those in prokaryotic cells. What's more, they have many membrane-bound organelles and a cytoskeleton to copy. In eukaryotes, cell division is achieved via a coordinated, repeating process called the cell cycle.

Right *Colored transmission electron micrograph of rod-shaped* Escherichia coli *(commonly known as E. coli) bacteria, which exist naturally in human intestines. In the center of the image, one bacterium is becoming two identical 'daughter' cells. The material colored yellow is the cells' DNA.*

VEGETATIVE REPRODUCTION

Some eukaryotic organisms make entirely new individuals by cell division. In plants, this is called vegetative reproduction. Bulbs and tubers, which provide a way of storing food for hard times, also provide the potential for a new plant. Stolons, or runners, are another way in which some plants produce offspring. They are stems that extend horizontally at ground level and have nodes from which a new plant can grow.

All the offspring produced by vegetative reproduction will be clones of the parent, because simple cell division results in cells that are genetically identical. Nevertheless, a multicellular organism, such as a plant, can still have many different types of tissue, each with its own characteristics and function. This is because cells differentiate into various types, thanks to different combinations of genes being expressed (switched on) in different cells. More on differentiation follows in chapter five.

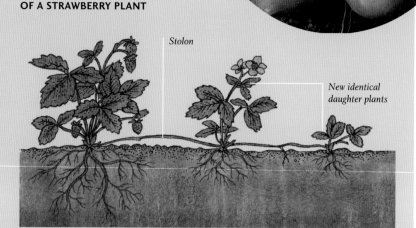

Right *A potato is a tuber and provides one way in which potato plants can reproduce, vegetatively. A whole new plant, identical to the original, can grow from a potato.*

ASEXUAL VEGETATIVE REPRODUCTION OF A STRAWBERRY PLANT

Stolon

New identical daughter plants

CELLS BEGET CELLS

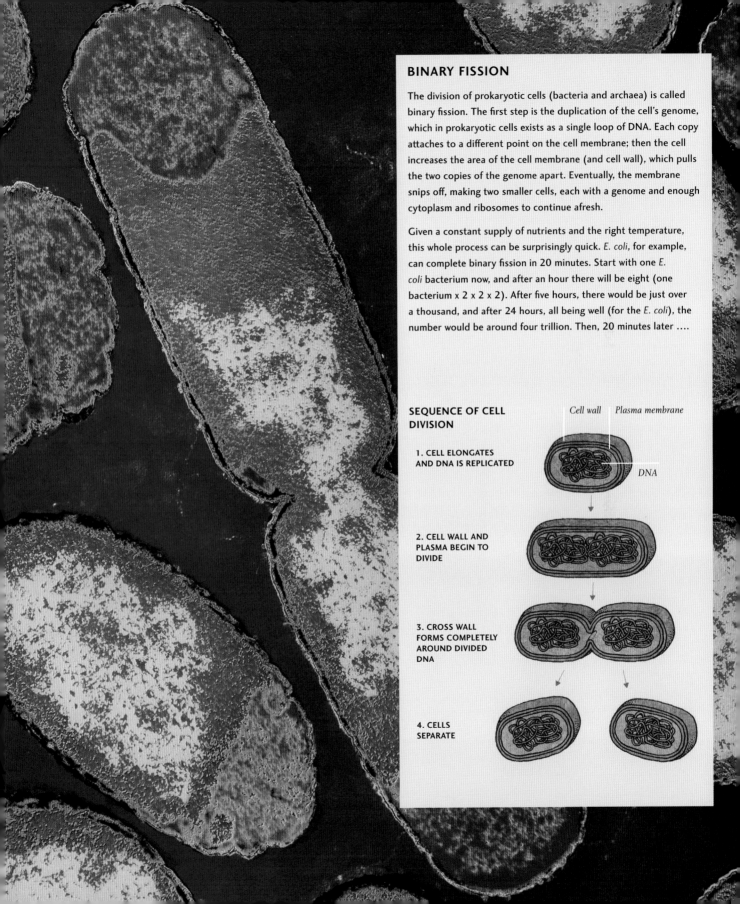

BINARY FISSION

The division of prokaryotic cells (bacteria and archaea) is called binary fission. The first step is the duplication of the cell's genome, which in prokaryotic cells exists as a single loop of DNA. Each copy attaches to a different point on the cell membrane; then the cell increases the area of the cell membrane (and cell wall), which pulls the two copies of the genome apart. Eventually, the membrane snips off, making two smaller cells, each with a genome and enough cytoplasm and ribosomes to continue afresh.

Given a constant supply of nutrients and the right temperature, this whole process can be surprisingly quick. *E. coli*, for example, can complete binary fission in 20 minutes. Start with one *E. coli* bacterium now, and after an hour there will be eight (one bacterium x 2 x 2 x 2). After five hours, there would be just over a thousand, and after 24 hours, all being well (for the *E. coli*), the number would be around four trillion. Then, 20 minutes later ….

SEQUENCE OF CELL DIVISION

1. CELL ELONGATES AND DNA IS REPLICATED
2. CELL WALL AND PLASMA BEGIN TO DIVIDE
3. CROSS WALL FORMS COMPLETELY AROUND DIVIDED DNA
4. CELLS SEPARATE

Cell wall   *Plasma membrane*   *DNA*

THE CELL CYCLE

The repeating chain of events by which eukaryotic cells proliferate begins when new cells have just been created.

1

G1 Phase: This is the period of time between the creation of a new cell and the copying of the nucleus. The G stands for gap, because cell biologists initially supposed that not much was going on. In fact, this is a very busy time. The cell has to grow so that it has enough of its vital material (proteins, RNA, organelles, cytoskeleton) to make two cells. And it has to prepare everything needed to copy the all-important DNA in the nucleus. So, for some, the G stands for growth.

2

G0: The G1 phase can be halted by what's going on outside the cell. Certain chemical signals from other cells, for example, can cause the cell to put the cycle on pause. Mature, quiescent cells, such as neurons in the brain, are stuck here and never go farther. Similarly, chemicals outside the cell can kick-start the cell cycle, moving the cell on to the G1 checkpoint.

3

G1 checkpoint: The cell will not proceed any farther unless certain conditions are met. If there is any damaged DNA, for example, certain enzymes flag it up and other enzymes fix it. Once all is well, the cell is committed to the next phase.

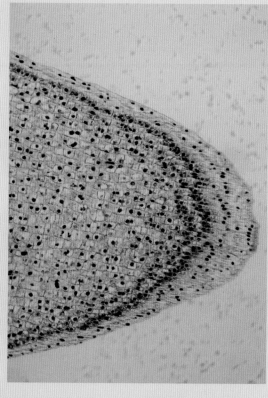

Above *Light micrograph showing a cross section of a corn root tip—a site at which cell division is very active. The cells' nuclei are stained dark blue.*

4

S Phase: Now the cell can begin manufacturing a copy of its genome (all its DNA); the S stands for synthesis. Different sets of enzymes come together to initiate then carry out the copying, to check its accuracy, and to make sure only one copy is made. Nearly all eukaryotic cells have more than one chromosome, so this is an enormously complicated process.

5

G2 Phase: Here is another gap in the cell cycle— and it, too, is anything but a peaceful break in proceedings. There's more frenetic growth with a lot of protein-building going on—in particular, the microtubules of the cytoskeleton, which will have an important role in the next phase.

6

G2 checkpoint: Again, the cell will not continue to the next phase until it has checked that any errors in the DNA copying have been corrected.

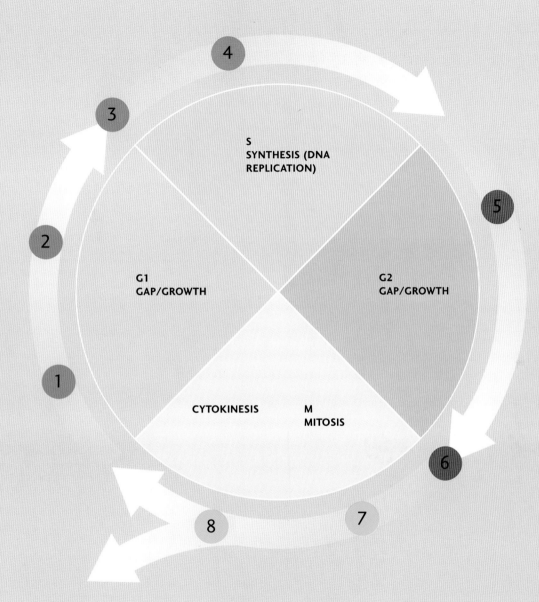

7

M Phase: The M is for mitosis. During this phase, the membrane around the nucleus disintegrates, then the microtubules of the cytoskeleton grab hold of the chromosomes and drag them to opposite sides of the cell's midline. At this stage there is, of course, another checkpoint, which ensures that the microtubules are properly attached. Immediately after mitosis is finished, the cell physically divides.

8

Cytokinesis: The physical separation of the two daughter cells is called cytokinesis. It is driven by microtubules pushing apart. In plant cells, microtubules left over from that process form a flat structure across the cell's midline, on which two new cell walls will be built. In both plant and animal cells, the much thinner microfilaments form a ring around the cell's midline and contract, snipping the cell in two and closing the membrane around both halves.

Once the two new daughter cells have been made, each one starts the cell cycle again, resulting in four granddaughter cells. Then each granddaughter cell will begin the cycle again ….

CELLS BEGET CELLS

Doubling up

The two G (gap) phases of the (eukaryotic) cell cycle are dominated by the remarkable, enzyme-driven construction procedure of protein building explained in chapter two. A similar process, equally remarkable, intricate, and important, is the duplication of DNA in the S phase of the cell cycle.

In pairs

In eukaryotic cells, the entire genome is typically billions of base pairs long and divided into several chromosomes, each a separate strand and all contained within the nucleus. Eukaryotic chromosomes come in pairs, because eukaryotic cells inherit one complete set of chromosomes from each parent (more on parents to follow). Kangaroo cells have 12 chromosomes (six pairs), humans have 46 (23 pairs), and pineapples 50 (25 pairs). The whole collection of chromosomes, arranged in their homologous pairs, makes up a family portrait known as a karyotype. The two members of each pair contain the same array of genes in the same order.

However, there may be different versions, or alleles, of each gene. Where two different alleles are present on the two paired chromosomes within a cell, one may dominate, so that the other gene remains inactive, or recessive. This explains why some inherited diseases can skip several generations, becoming a problem only when an individual is created with the defective allele on both homologous chromosomes. Some genes are codominant, in which case both alleles will be expressed, leading to a genetic hybrid.

Copying the DNA

DNA replication begins at specific nucleotide sequences called replication origins; here, enzymes attach to initiate the process of copying. Remarkably, at these regions, there are many more bonds between the A and T nucleotides than normal; this bond is weaker than the C–G bond, so at a cluster of A–T bonds, the DNA unzips more easily than in other parts of the chromosome.

Below *Illustration of DNA replication, making two copies of DNA double helices where first was one. Duplication proceeds from two "duplication forks," formed by the enzyme helicase. The enzyme DNA polymerase adds new nucleotides.*

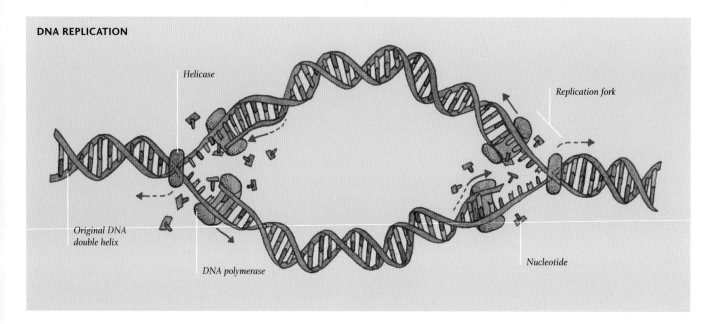

DNA REPLICATION

Helicase

Replication fork

Original DNA double helix

DNA polymerase

Nucleotide

Another enzyme now attaches and begins to unwind and unzip the DNA double helix into its two constituent strands. Yet more different enzymes now begin the process of replication, attaching new A, G, C, and T nucleotides in the correct sequence, mirroring each strand. Eventually, there are two identical, double-stranded copies of each chromosome. Notice that one strand in each was present in the original double helix, so it is not the case that one chromosome is the original and one is a copy; both are equivalent duplicates. Just so they do not get lost or mixed up, they remain attached, in pairs, glued together by a protein called cohesin. These "sister chromatids" will only separate just before the cell divides.

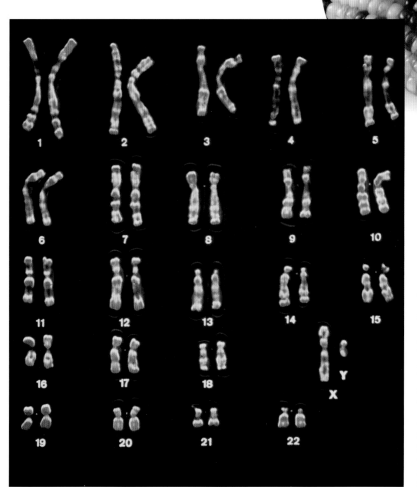

Below *Cob of a genetic hybrid corn plant—two different versions of a gene that codes for a pigment protein are both active.*

Left *False color image of a set of human chromosomes. There are two of each chromosome—46 in total. The chromosomes within each pair are homologous, not identical: one is from the individual's mother, one from the father. Sex is determined by a permutation of two sex chromosomes, named X and Y in humans. A female has two X chromosomes, a male has an X and a Y. Although they are homologous, they are very different with completely different sets of genes along their length.*

Mitosis

Perhaps the most dramatic event of the cell cycle is mitosis (M phase). In this, the chromosomes condense to become visible, threadlike packages and then separate into two sets of duplicate chromosomes.

One makes two

The duplicated chromosomes remain stuck together through the G2 phase, while the cell prepares for mitosis, the most spectacular part of the cell cycle. Although it is one continuous chain of events, biologists divide it into subphases: prophase, prometaphase, metaphase, anaphase, and telophase. The prefixes here don't actually have anything to do with what's going on: pro- means "before," meta- means "middle," ana- means "onward," and telos- means "end."

In prophase, the chromosomes, attached like conjoined twins, change from long and stringy to tightly packed. These remarkable structures are now sufficiently condensed to be visible in an ordinary light microscope. Each one appears as an X because the two duplicates—the sister chromatids—are held together in one place, normally fairly near the center. There are two versions of each X-shape structure, because there were two versions of each chromosome before duplication. Each pair is about to be ripped apart. Toward the end of this phase of mitosis (in all but plant cells, which do things slightly differently), tiny structures called centrosomes become active. They are centers on which microtubules are built, and those fast-growing microtubules are about to have a very important role. The two centrosomes move so that they are on opposite sides of the nucleus.

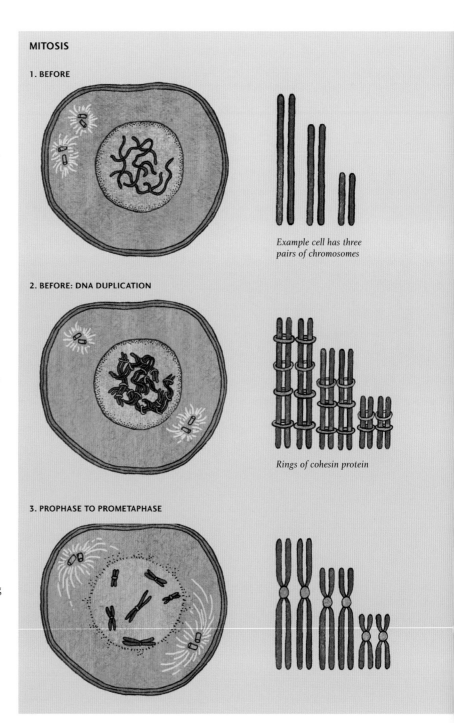

MITOSIS

1. BEFORE

Example cell has three pairs of chromosomes

2. BEFORE: DNA DUPLICATION

Rings of cohesin protein

3. PROPHASE TO PROMETAPHASE

CELLS BEGET CELLS

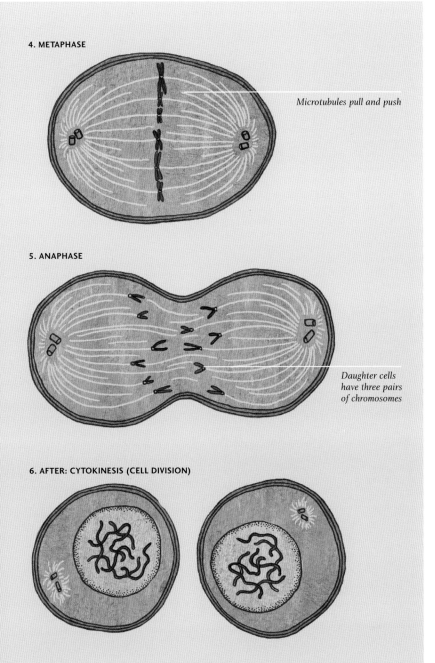

Microtubules pull and push

Daughter cells have three pairs of chromosomes

Pull and push

Now things start to get serious. During prometaphase, the membrane around the nucleus disintegrates, leaving the chromosomes wide open and ready to be pulled apart. To that end, during metaphase, microtubules growing out from the centrosomes attach to either side of each X-shape conjoined pair of duplicate chromosomes, and a remarkable tug-of-war aligns them all along the cell's midline. After all this crazy activity, it is reasonable that the cell needs to make sure everything is OK. If any of the microtubules are incorrectly fixed to a chromosome, or if the chromosomes are not properly aligned across the middle of the cell, the whole process stops until it is corrected.

Anaphase is the most dramatic part of mitosis. The microtubules holding onto the middles of chromosomes now begin shrinking. Each sister chromatid has one microtubule holding it, and the two microtubules on each chromosome are from opposite sides of the cell. The consequence is that the conjoined twins are ripped apart, the happy result of this destruction being that one copy of each chromosome is pulled to each side of the cell. The cell now contains two complete, separated, identical genomes.

During anaphase and into telophase, the microtubules that are not attached to the chromosomes join in the center of the cell and push against each other, elongating the cell into an oval shape. All that remains in telophase is to clean up the mess created by these incredible happenings and close the membrane off. Now the daughter cells are complete and ready to start the whole cycle again.

CELLS BEGET CELLS

MITOSIS GALLERY

Mitosis is the most important phase of the cell cycle, and the most visually wondrous. The sequence of images are light micrographs of mitosis in cells from a rat kangaroo's kidney. Fluorescent dyes have stained DNA blue and microtubules green. Feast your eyes and dazzle your brain.

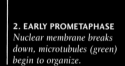

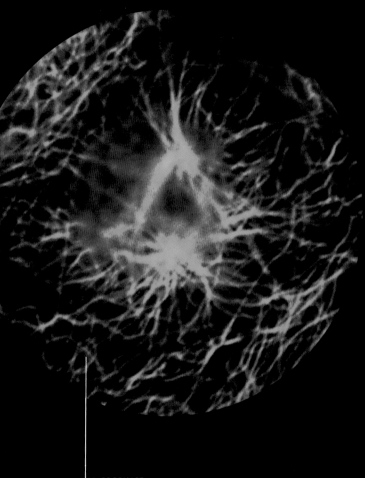

2. EARLY PROMETAPHASE
Nuclear membrane breaks down, microtubules (green) begin to organize.

3. PROMETAPHASE
Microtubules begin to pull the chromosomes, lining them up ready for separation into two identical sets.

1. PROPHASE
All chromosomes (blue) still in the nucleus, but condensing.

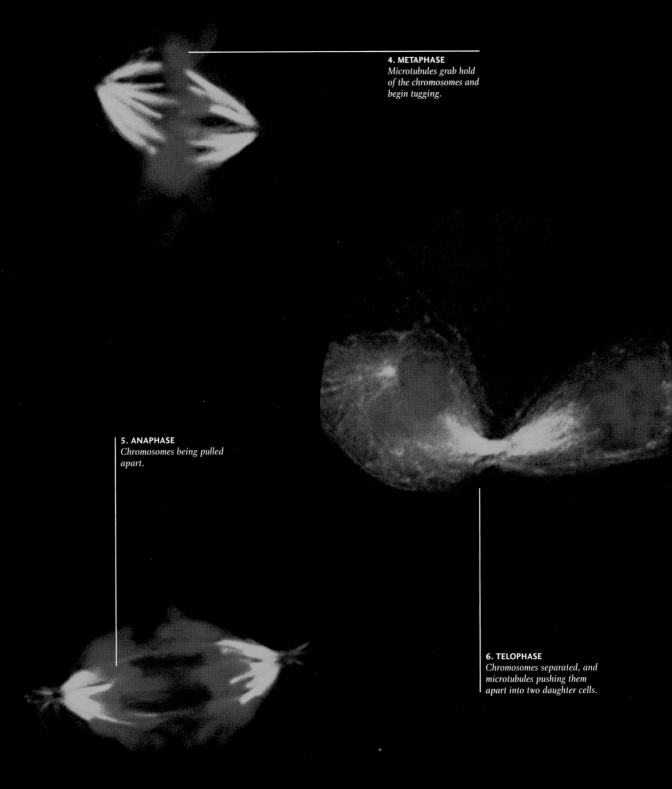

4. METAPHASE
Microtubules grab hold of the chromosomes and begin tugging.

5. ANAPHASE
Chromosomes being pulled apart.

6. TELOPHASE
Chromosomes separated, and microtubules pushing them apart into two daughter cells.

Sexual reproduction

The cell cycle and binary fission can make one cell only into two identical copies; the genome is passed on to daughter cells unchanged. There is another kind of cell division, however, which is at the heart of sexual reproduction. This process introduces genetic variation, guaranteeing that no two offspring are identical. Once again, it all comes down to dances of the chromosomes.

Two halves make a whole

Some eukaryotes reproduce asexually—by budding, for example (see box), or, in the case of plants, using vegetative growth to form a new individual (page 60). But all eukaryotes can reproduce sexually. Sex is the reason why the chromosomes in eukaryotic cells come in homologous pairs: one chromosome in each pair comes from the female parent, one from the male. Cells that contain a pair of each chromosome are described as diploid. All our somatic cells—the ordinary cells in our skin, brain, muscles, and so on—are diploid, with 46 chromosomes altogether. But there are some cells whose nuclei contain only one set of chromosomes. These haploid cells are the germline cells, called gametes. Our gametes contain only 23 chromosomes. They are made for pairing.

It is the union of two haploid gametes, each with one set of chromosomes, that unite to become a diploid cell. Each chromosome in the male gamete (the sperm) corresponds to one in the female gamete (the egg); this is why they exist in homologous pairs inside our diploid somatic cells. A fertilized egg—or indeed any diploid cell created by the union of two haploid gametes—is called a zygote. And it is the zygote that divides, via the ordinary cell cycle, to begin the creation of a new, multicellular individual. Each of its somatic cells contains the same diploid genome. But, ultimately, some of its diploid cells divide in a different way to become the haploid gametes necessary for the next generation.

Doubling up

Unlike the grow-duplicate-divide-repeat cell cycle, the production of gametes is not a repeating chain of events. It has a beginning and an end. The process begins with one diploid precursor cell and ends with four haploid gamete cells, all unique. The chromosomes in the precursor cell are in pairs—but, importantly, the two members of each homologous pair are not identical to each other. They consist of the same genes, but there may be different versions of each. The first step toward creating gametes is the duplication of all the DNA in the nucleus, just as it happens in the (S phase of the) cell cycle. Once that is finished our precursor cell has two identical

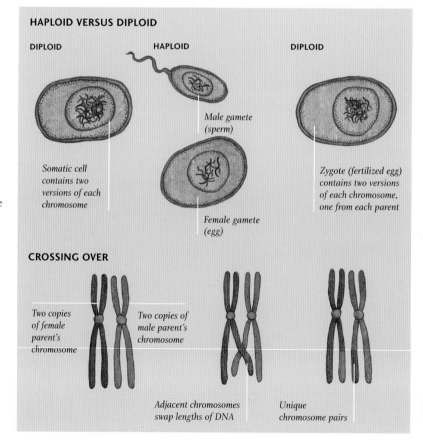

HAPLOID VERSUS DIPLOID

DIPLOID — Somatic cell contains two versions of each chromosome

HAPLOID — Male gamete (sperm); Female gamete (egg)

DIPLOID — Zygote (fertilized egg) contains two versions of each chromosome, one from each parent

CROSSING OVER

Two copies of female parent's chromosome

Two copies of male parent's chromosome

Adjacent chromosomes swap lengths of DNA

Unique chromosome pairs

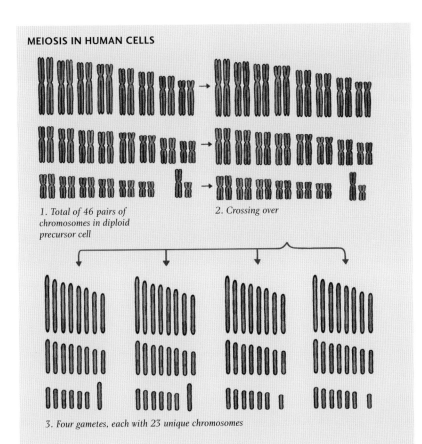

MEIOSIS IN HUMAN CELLS

1. Total of 46 pairs of chromosomes in diploid precursor cell
2. Crossing over
3. Four gametes, each with 23 unique chromosomes

copies of each chromosome. The two copies—the sister chromatids—stick together like conjoined twins in an X shape.

So far everything is as it happens in the cell cycle. But here is where things start to happen differently. The chain of events from here on is called meiosis. Incredibly—and no one knows how—each X-shape pair of conjoined twins finds and attaches itself to its equivalent. The homologous pairs of duplicate chromosomes—two similar pairs of conjoined identical twins—now stick together. There are four chromosomes in each bundle altogether. At this point something amazing happens. Neighboring chromosomes within each four-chromosome bundle exchange sections of their DNA in a procedure called crossing over. The result is a mixing of the genomes of this individual's mother and the father. This is one of two steps that injects variation into the gametes-to-be.

Half and half again

Our single cell now has a nucleus with sets of four homologous chromosomes, multiplied by however many chromosomes there are in the complete set. Thanks to crossing over, not one chromosome is identical to any other. The homologous ones are still clinging together in four-chromosome bundles—for now. The rest of the process consists of two rounds of cell division, after which the number of each kind of chromosome will have gone from four in one cell to one in each of four cells. There is one more feature of meiosis that injects variation into the gametes. When the chromosome twin pairs line up, ready to be pulled apart by the microtubules, they do so in a random arrangement. The chromosomes—already mixed up along their length, thanks to the crossing over, are mixed even more.

After the two rounds of cell division, the required four haploid cells are ready for action. Each one is able to fuse with another gamete that has been produced in the same way in an individual of the opposite sex of the same species.

BUDDING YEASTS

Yeasts are single-celled eukaryotes. Most of the time they reproduce asexually, by budding—something similar to cell division. In that case, the offspring are identical to their parents. However, yeasts can also live in a form in which they have only one set of chromosomes not two. These haploid cells merge their genomes—in other words, they reproduce sexually. The likelihood of a yeast entering into sexual reproduction is increased if the yeasts are put under any kind of stress—for example, from high temperatures or a harmful chemical environment. This increases variation in the offspring, increasing the chance that some of them will survive, because some new combination of genes may have a better chance of survival. In general, enhanced variation increases the rate of evolution.

MEIOSIS GALLERY

Sexual reproduction involves the union of two haploid gametes—and meiosis is the remarkable process by which those gametes are produced. This sequence shows highlights of meiosis in a lily plant producing pollen cells, the male gametes. Note that, although the images are arranged in a loop, meiosis is not a cycle.

1. PROPHASE I
Chromosomes (purple) have already been duplicated, so there are now 48 chromosomes (24 homologous pairs), instead of the normal 24 (12 homologous pairs). The duplicated chromosomes now undergo crossing over and random allocation.

5. METAPHASE II
In the second phase of meiosis, the two new nuclei now break down and the chromosomes divide at the same time to form four groups, aligned along a new equator.

6. TELOPHASE II
Microtubules inside the two cells pull the chromosomes apart and two cells become four, as a new membrane and cell wall form. Each of the four cells is a haploid gamete.

CELLS BEGET CELLS

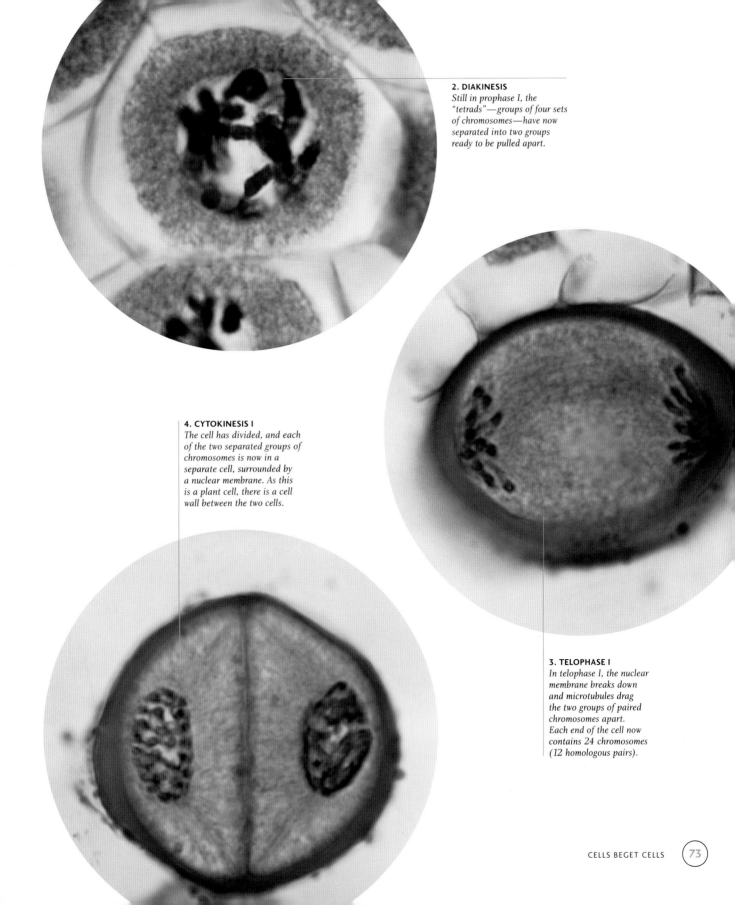

2. DIAKINESIS
Still in prophase I, the "tetrads"—groups of four sets of chromosomes—have now separated into two groups ready to be pulled apart.

4. CYTOKINESIS I
The cell has divided, and each of the two separated groups of chromosomes is now in a separate cell, surrounded by a nuclear membrane. As this is a plant cell, there is a cell wall between the two cells.

3. TELOPHASE I
In telophase I, the nuclear membrane breaks down and microtubules drag the two groups of paired chromosomes apart. Each end of the cell now contains 24 chromosomes (12 homologous pairs).

CELLS BEGET CELLS

Gamete galleries

The gametes produced by meiosis in eukaryotic organisms come in many shapes and sizes, and they join together in a bewildering variety of different ways. But in every case two haploid cells, created by meiosis, come together to form a diploid cell that is the beginning of a new individual.

Animal gametes

In animals, the gametes are the sperm and the ovum (egg cell). Both are produced by meiosis, but they are very different. Four sperms are produced for every one precursor cell: classic meiosis. The sperm cells do not need to live long—just long enough to find and fertilize an egg. So they do not need to

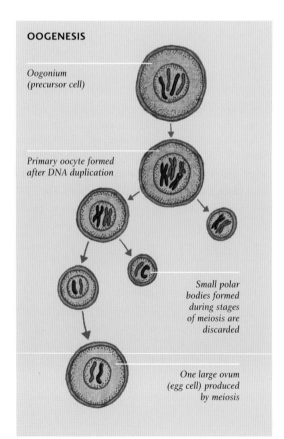

OOGENESIS

Oogonium (precursor cell)

Primary oocyte formed after DNA duplication

Small polar bodies formed during stages of meiosis are discarded

One large ovum (egg cell) produced by meiosis

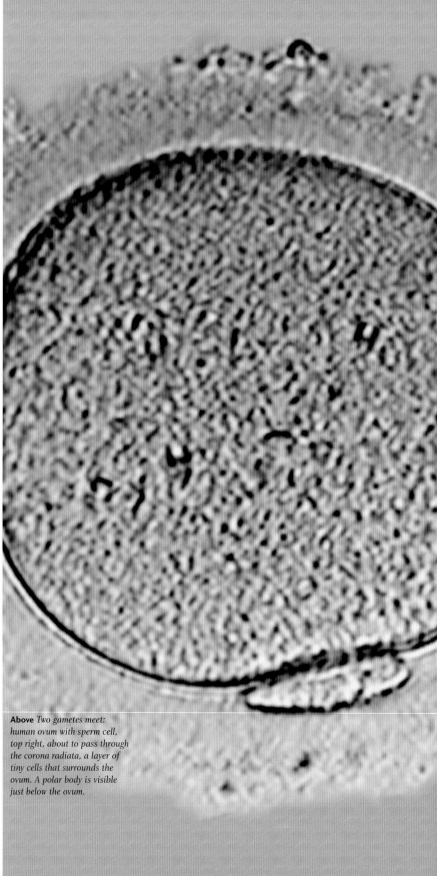

Above *Two gametes meet: human ovum with sperm cell, top right, about to pass through the corona radiata, a layer of tiny cells that surrounds the ovum. A polar body is visible just below the ovum.*

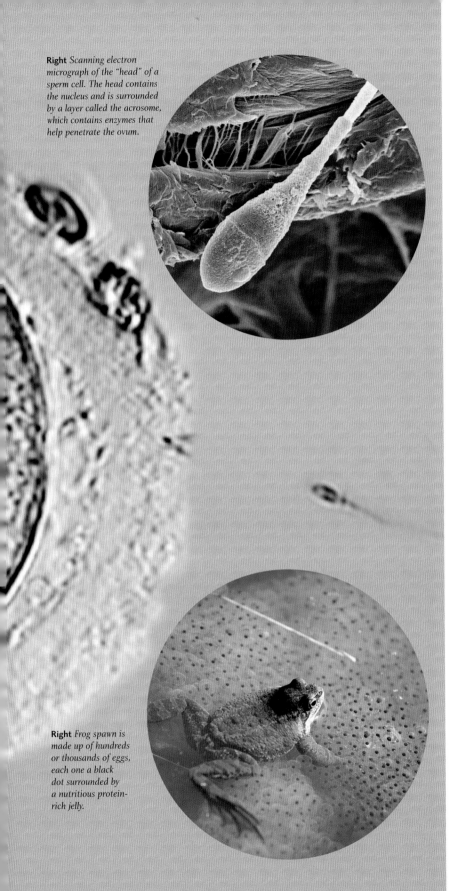

Right Scanning electron micrograph of the "head" of a sperm cell. The head contains the nucleus and is surrounded by a layer called the acrosome, which contains enzymes that help penetrate the ovum.

Right Frog spawn is made up of hundreds or thousands of eggs, each one a black dot surrounded by a nutritious protein-rich jelly.

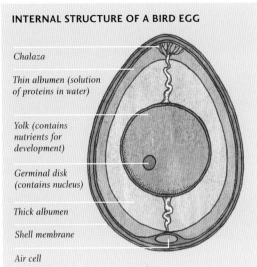

INTERNAL STRUCTURE OF A BIRD EGG

- Chalaza
- Thin albumen (solution of proteins in water)
- Yolk (contains nutrients for development)
- Germinal disk (contains nucleus)
- Thick albumen
- Shell membrane
- Air cell

carry with them much in the way of sustenance, which also means they can be small and mobile. Things are different for the egg cell. The ovum has to contain enough nutrients to begin initial development of a new individual after fertilization, so all ova are relatively large cells.

In birds, reptiles, fish, amphibians, and insects, that initial post-fertilization growth takes place outside the body. Consequently, the eggs are considerably larger than they are in mammals, where development takes place inside the female parent's body. The development of egg cells is put on pause at several stages to allow for the necessary growth. One crucial point at which the egg's development pauses is just after the precursor cell's DNA has been duplicated. At that time, there are two copies of each gene, so the cell can double the rate of protein manufacture. Only one ovum is produced for each precursor cell, not the four that would normally be the result of meiosis. This is achieved by the creation of polar bodies, small cells created in each round of cell division during meiosis. Each polar body dies off before it can divide. Because the single resulting cell has so much invested in it, eggs typically mature one at a time, when mating is about to occur.

CELLS BEGET CELLS

Left *Close-up of moss: The main part of the plant is gametophyte, made of haploid gametes. The tall structures are sporophytes, made of diploid cells.*

Plant and fungal gametes

Many plants normally reproduce asexually. Bulbs, rhizomes, and stolons, for example, can produce new individuals purely by mitosis. Of course, the offspring produced in this way are all genetically identical. All plants can also reproduce sexually. The sexual life cycle of plants is called the alternation of generations, and it differs from that of animals. The gametes, produced by meiosis, undergo repeated division by mitosis; in many cases, they proliferate to form a significant, multicellular part of the plant. This structure is called a gametophyte. A green carpet of moss, for example, is all gametophyte. Each moss plant is made up entirely of haploid gamete cells that have been created from a single gamete. There are male and there are female moss plants. The equivalent in animals would be a whole body made of egg or sperm cells. In conifers, the gametophyte is tiny, hidden inside the cones—there are male and female types. In flowering plants, the tiny male gametophyte is hidden inside pollen grains, while the female gametophyte is the ovule.

Gamete cells from two gametophytes unite in fertilization, and the resulting diploid zygotes proliferate, producing structures called sporophytes. Each sporophyte is a multicellular organ made of diploid cells. In mosses, the (diploid) sporophyte is lifted above the rest of the (haploid) plant on a stalk. In the conifer and the flowering plant, the sporophyte is the dominant part—all the leaves and stems are diploid. Some of the cells of the sporophyte undergo meiosis, producing haploid spores, which divide by mitosis to form the gametophytes. And around it goes again.

Flowering plants carry out a unique procedure called double fertilization, which has an added benefit: it provides food for the new individual in a package called the endoplasm. Inside a pollen grain is one male gametophyte that divides to form two sperm cells as it reaches the ovule. Inside the ovule are eight identical haploid nuclei, most of them surrounded by a cell wall. One of the surrounded nuclei is the ovum, and one of the two sperms fertilizes it, forming the beginning of a new diploid plant. Two of the eight (female) nuclei inside the ovule are not surrounded by a cell wall. The second sperm cell fertilizes both these nuclei to become a triploid cell. That cell divides repeatedly, forming the multicellular endoplasm.

Fungi mostly reproduce asexually—break off a piece of a fungus's body, and it can generate a new organism. But fungal bodies are mostly haploid cells, and when these cells meet haploid cells from another individual the two cells merge. The two cells' nuclei—and their genomes—can stay separate within each cell for very long periods. Eventually, the two genomes do fuse, and they produce a new diploid spore-producing body and a new, unique individual.

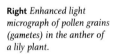

Right *Enhanced light micrograph of pollen grains (gametes) in the anther of a lily plant.*

Far right *Light micrograph of (stained) cells in bread mold, Rhizopus nigricans, undergoing conjugation, a process in which haploid cells come together to form diploid spores.*

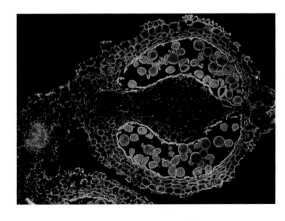
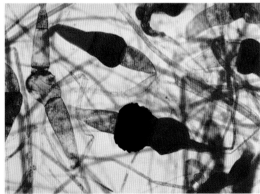

Variations on a theme

Sexual reproduction mixes up genes, creating a constantly shifting array of combinations. But there are other ways in which genetic variation is introduced—including the creation of brand new genes.

Genetic typos

A genome is a very long sequence of letters—the As, Cs, Gs, and Ts of the nucleotides that make up DNA. During duplication of the genome, the nucleotides are added one after another in rapid succession from a reservoir of nucleotides in the cytoplasm. Sometimes the wrong one is selected. It is like a very experienced typist copying something out quickly and making the occasional error. A typist would have an editor to check the work and correct the errors, and cells have sophisticated inbuilt mechanisms that do the same. However, as occasionally happens with typists and editors, some mistakes still make it it through ….

In the cell, changes can also be introduced when the genome is not being copied. These changes can be the result of radiation or chemical reactions affecting the bonds between atoms in the DNA molecule.

Down the line

Changes to DNA are known as mutations. If they occur within genes, they may be damaging, and can result in a cell's death. In a multicellular organism, a harmful mutation that occurs in just one cell will normally have little effect—although it may lead to cancer (see chapter six). Of course, a damaging mutation in a single-celled organism spells that organism's end. However, not all mutations are life-threatening, even to a single cell.

When a mutation does not spell the end of a cell it is duplicated along with the rest of the DNA, becoming part of the genome of the daughter cell; and that cell's daughter cells. Most of the cells of a multicellular organism are somatic cells, and they carry out their functions for the life of the organism (or for part of it), and they die when that organism dies (or before). Any mutations in the genomes of somatic cells are not carried on to the next generation of the organism as a whole. However, a nonlife-threatening mutation in gametes does have the potential to be passed on to the next generation. The same is true of any nonlife-threatening mutation in a single-celled organism. The most common mutation changes just one nucleotide. These single-nucleotide polymorphisms can result in heritable diseases, such as sickle cell disease (see box below).

Any mutations passed on to the next generation are a source of variation among a species. Some mutations will have no noticeable effect; some will have a huge effect. Some will prevent a particular protein from being produced or change the protein very slightly. Sometimes a short section of DNA will be lost, and two genes will fuse together. In prokaryotic cells variation can also be introduced by a phenomenon known as horizontal gene transfer (see box on facing page).

SINGLE-NUCLEOTIDE POLYMORPHISMS

Sickle cell disease is caused by a mutation on chromosome 11, in the gene for the protein haemoglobin. The mutation is a single-nucleotide polymorphism (SNP)—a simple substitution of one base in place of another (in this case, T for A)—that affects the protein's shape. The misshapen protein makes it more difficult for red blood cells to move easily through capillaries. The mutation is passed from generation to generation.

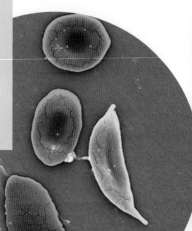

Below *Light micrograph showing three normal red blood cell and one "sickle" red blood cell. The latter contains a faulty hemoglobin protein.*

HORIZONTAL GENE TRANSFER

Genetic material is normally passed vertically from generation to generation. But it is also possible for prokaryotic cells (and, to a far lesser extent, eukaryotic cells) to transfer genetic material among themselves within the same generation. There are three ways in which this horizontal gene transfer happens. First, if a prokaryotic cell dies and its contents spills out, DNA can be taken up by a nearby prokaryotic cell and become part of that cell's genome. Second, a virus can transfer sections of DNA from one cell to another.

The third way involves transferring a piece of DNA directly between two living cells. In addition to their main, single genomic chromosome, most prokaryotic cells also have smaller loops of DNA called plasmids, which code for specific proteins—typically proteins that help protect the cell. One cell can extend a thin appendage called a pilus from its membrane to another cell, and a copy of a plasmid passes across from one cell to the other.

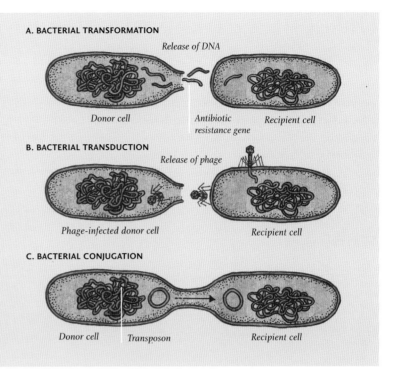

Copy and paste

There is one additional source of variation: sequences of DNA outside genes that have a tendency to copy (or cut) and paste themselves to different points within the genome. These transposable, or movable, elements are called transposons. They can be pasted right in the middle of a gene, or they can simply cause two adjacent genes to move relative to each other. Either change will have an effect on the genome overall and, therefore, on the individual that owns it.

The founder of the science of genetics, Gregor Mendel, worked with pea plants in the 1860s. One of the traits he studied was how wrinkled or smooth the peas were (see page 20). Modern geneticists now know that the difference between the two types is the effect of a transposon. In the plants that produce wrinkled peas, the gene for a particular protein no longer works because a transposon was pasted into it thousands of years ago. The damaged gene is nonlife-threatening, so it is carried through the generations. However, if a plant's genome has at least one copy of the original gene (on one of the pair of homologous chromosomes that contain the gene), the protein will be produced and the pea will be smooth. In other words, the "round" version is dominant, because it successfully codes for the protein that makes the pea smooth. The peas will be wrinkled only if both versions of the (recessive) gene are the mutated allele.

The fact that the characteristics of an organism are coded in DNA, passed down the generations, and able to mutate is the basis for one of the natural world's most fundamental processes: evolution.

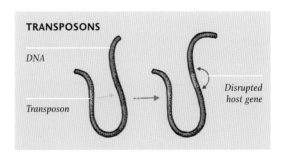

Descent with modification

Genetic variation is introduced into DNA by mutations and by transposons, and it is enhanced in the production of gametes and by their union in eukaryote sexual reproduction, as well as by horizontal gene transfer in prokaryotes. Variation is an essential part of the process of evolution. Without changes to DNA, all cells would have been identical since time immemorial.

A theory of change
In his 1859 book *On the Origin of Species*, British naturalist Charles Darwin described the basis of his theory of evolution as "descent with modification."

THE POWER OF SELECTION

Mutations and other genetic variation can survive only if the individuals carrying them manage to reproduce. Darwin noted that selective breeding could introduce dramatic changes in the appearance of domesticated plants and animals. He experimented with selective breeding in pigeons, mating together birds that displayed particular traits to see if he could sustain and enhance those characteristics. He called this "artificial selection," because he realized that similar changes are brought about by natural selection in the wild.

There are a multitude of selective pressures in the environment—such as changing habitats, dwindling food sources and competing species—and they change over time. The organisms that will most probably pass on their genes are the "fittest" ones, those best suited to their environment. Darwin noticed distinct differences between mockingbirds on various islands of the Galápagos and realized that each one was adapted to the particular conditions on the island where it lived. Change is inevitable; natural selection is the mechanism that inadvertently steers its course.

Above *It was these mockingbirds and others like them, which Charles Darwin collected in the Galápagos Islands, that first gave him his key insight into natural selection.*

Above *Darwin's own sketch of the 'tree of life', which 'fills the crust of the earth with its dead and broken branches, and covers the surface with its ever-branching and beautiful ramifications'.*

Below *Sedimentary rocks from the 3.48 billion-year-old Dresser Formation, in Western Australia. The structures in the rocks were produced by interactions of the sediment with early single-celled organisms.*

He was referring to the fact that the characteristics of one living thing are passed to its descendants—and that those characteristics are apt to change. It would have pleased him to learn that the basis for both descent and modification is the behavior of one remarkable chemical compound found in all cells.

The entirety of an organism's DNA—its genome—contains all the information necessary to build that organism. The genes coded into the DNA determine the organism's characteristics, and genes are passed down through generations. The differences between any two individuals, whether they belong to the same species or not, is down to genetic variation—differences in their genes. Small changes in genomes within a population can accumulate to such a degree that, over long periods of time, entire species die out and new species emerge. Of course, not all changes survive—the essential ingredient that makes Darwin's theory complete is "selection" (see box on facing page).

The tree of life

A key idea that Darwin introduced is the notion of common descent. He realized that all living things are related—as if all of nature is like a tree, and all species living today are at the ends of branches that are still growing and dividing. Further back along the branches there will be a point where a branch diverged. In the same way, if we look back far enough into a species' past, we will find two modern species have a common ancestor. Continuing back along the branches eventually the trunk is reached from which all the branches grew. In the same way, if we go back far enough, we will find every organism alive today has a common ancestor with every other one. No one knows what that common ancestor was like, or how the first cells came to be, but it happened at least 3.5 billion years ago (see box below).

There is a good deal of evidence within cells to support the idea of common descent—not least the fact that at the molecular level, living things are remarkably similar to each other. The very fact that every living thing uses DNA as the basis of inheritance and protein manufacture, and ATP as the energy currency, for example, is evidence in favor of common descent. But the best evidence comes from detailed, comparative studies of the genomes of different species.

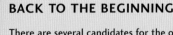

BACK TO THE BEGINNING

There are several candidates for the oldest evidence of life on Earth. All are around 3.5 billion years old. Some are microfossils of communities of simple cells; some are simply traces of biochemical activity in ancient rocks. Despite the ubiquity of DNA in living things today, there is no evidence that it was involved in these very early life forms. However, it is probable that early life involved nucleic acids in some way. Nucleobases—of confirmed extraterrestrial origin—have been found in meteorites. Ground-based experiments suggest that these probably formed in clouds of water ice around star-forming regions in space. It is known that Earth's oceans formed from water deposited by comets impacting the planet—it is probable that these comets brought with them the essential ingredients for life to begin, ready to go.

CELLS BEGET CELLS

Know your relatives

Thousands of individuals across hundreds of species have had their entire genomes sequenced since the late 1990s. Many humans' genomes are among them, but the majority of those sequenced are bacteria, because they have relatively simple—and also small—genomes. However, geneticists have gleaned detailed knowledge of gene sequences in many thousands more species. Such knowledge can be used to work out how closely related any two individuals are. Within a single species, relatedness can be measured by nucleotide diversity—the number of differences in actual nucleotides along the DNA sequence. The nucleotide diversity of humans is around 0.2 percent, so no more than about one in 500 nucleotides are different across all humans.

It is possible to measure the relatedness between different species in a similar way, looking for genes that are present in two genomes and how closely the genes resemble each other. Chimpanzees are our closest relative; we share more than 96 percent of our DNA sequence (although that figure takes into account the fact that some sequences are repeated in one species and not in the other, and the "real" figure is higher). Evidence from genome studies and fossil finds show that humans and chimpanzees share a common ancestor that lived between 4 and 6 million years ago. We have a great deal in common with all the great apes (see box). Similarly we share around 90 percent of our DNA with cats, about 80 percent with cows, 60 per cent with fruit flies, and even 50 percent with bananas. This last figure is remarkable, given that plants and animals have been on separate branches of the tree of life for more than a billion years.

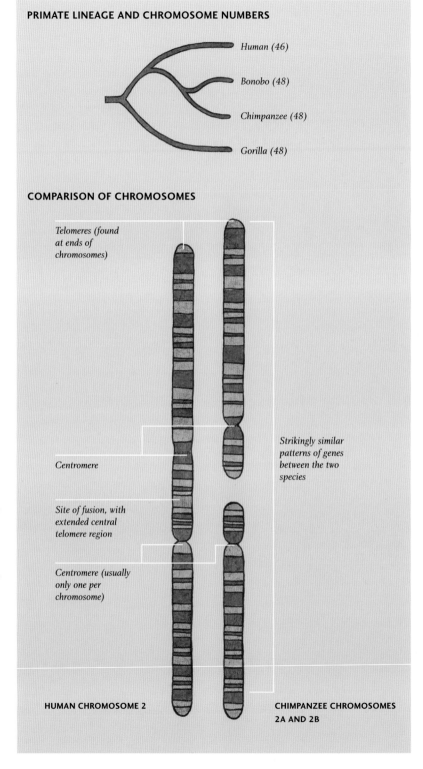

PRIMATE LINEAGE AND CHROMOSOME NUMBERS

Human (46)
Bonobo (48)
Chimpanzee (48)
Gorilla (48)

COMPARISON OF CHROMOSOMES

Telomeres (found at ends of chromosomes)

Centromere

Site of fusion, with extended central telomere region

Centromere (usually only one per chromosome)

Strikingly similar patterns of genes between the two species

HUMAN CHROMOSOME 2

CHIMPANZEE CHROMOSOMES 2A AND 2B

OUR COUSINS THE CHIMPS

By studying physiology and embryology, Darwin realized that humans have a great deal in common with apes and that our species and the other great apes probably evolved from a relatively recent common ancestor. Comparative genome studies back this theory remarkably well. One curious fact, which seems at first to contradict our close relationship with other apes, is the fact that humans have 23 pairs of chromosomes, while all other apes have 24. Studying the two sets of chromosomes more closely, however, it turns out that human chromosome 2 has all the same genes, in the same order, as two ape chromosomes joined together. Also, human chromosome 2 has four regions that were once clearly the ends of two chromosomes. Finally, it has two centromeres—these are regions where microtubules attach and pull duplicate chromosomes apart during cell division. Normal chromosomes have one each.

Working with the code

The combination of DNA mutations and natural selection creates remarkably diverse adaptations and species interrelationships throughout the natural world. But it is a very slow and ultimately a random process. Tinkering directly with genomes, engineering or modifying them, has opened up a whole new world of possibilities.

What about us?

Thanks to evolution, DNA develops genes that ensure the organisms that carry it have the right characteristics to help it survive in a challenging and changing world. Yet DNA does not do this on purpose. It is just that if it didn't work that way, life would not have survived. The changes that persist are only guaranteed to improve the chances of the DNA being copied; evolution takes no notice of our needs—and why should it? So, for centuries, human beings have been using artificial selection, for example, to breed improved varieties of plants and animals—selecting for increased resistance to pests, improved yields, or better flavors. Perhaps the most dramatic effect of these efforts can be seen in what came to be known as the "Green Revolution," which began in the 1950s and improved wheat yields so greatly that it averted the threat of mass famine (see box below).

Since the 1970s, there has been a quicker, more direct, and more flexible way in which we can bend nature's genomes to our needs: genetic engineering. What is more, its central feature—transferring genes from one species to another— is something that does not happen in nature. It is possible because all species' genomes are based on DNA, and all use the same code.

Below *India is one of the countries to embrace biotechnology. The successful collaboration of scientists from 15 nations has resulted in the sequencing of the wheat genome, which will enable plant researchers to breed higher-yielding varieties of a grain on which almost a third of the world population relies.*

THE GREEN REVOLUTION

In the 1950s, American geneticist Norman Borlaug utilized careful cross-breeding to produce a high-yield, disease-resistant strain of wheat. The key was to breed a short, stout stalk that was strong enough to support the increased weight of the grains. This new strain was planted in several countries in which mass famines had been experienced or predicted. Together with improved infrastructure and increased use of pesticides, the new wheat strain averted famines and helped to sustain the world's burgeoning population. Whether that's a good thing or not is debatable, but Borlaug is generally lauded as a hero who saved a billion lives, and the huge influence of the wheat strain he created is testament to how widespread an effect small changes to molecules of DNA can have.

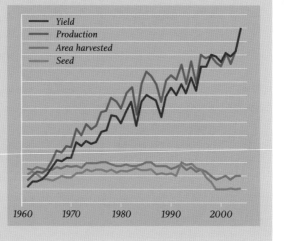

TOTAL WORLD PRODUCTION OF COARSE GRAIN
1961–2004
Source: FAO

— Yield
— Production
— Area harvested
— Seed

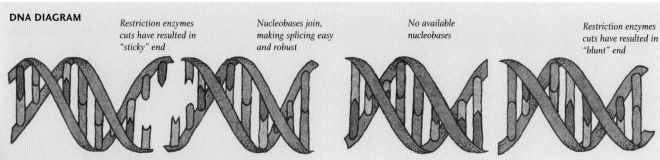

DNA DIAGRAM

Restriction enzymes cuts have resulted in "sticky" end

Nucleobases join, making splicing easy and robust

No available nucleobases

Restriction enzymes cuts have resulted in "blunt" end

STICKY ENDS

BLUNT ENDS

Genetic engineering

The crucial enabling step in the development of genetic engineering was the discovery of restriction enzymes in the 1960s. Restriction enzymes recognize particular short sequences of nucleotides, latch on to DNA at those locations, and then cut through the DNA like a pair of scissors. They are produced naturally in bacteria and probably evolved as a protection against attack by viruses. More than a hundred different restriction enzymes are known, each one of which cuts the DNA molecule at a different nucleotide sequence. Some cut straight across the double helix, leaving the DNA molecule with "blunt ends"; others cut in a dogleg shape, leaving "sticky ends" that enable pieces of DNA to join together more easily.

Restriction enzymes are used in the first step of genetic engineering to cut out specific target genes from an organism—genes that can produce a beneficial effect when inserted into the cells of another organism. The sticky ends of cut pieces of DNA are particularly useful when the pieces are joined onto DNA inside the cells of another organism. There are many different ways in which the cutout section of DNA can be inserted into a recipient organism—normally it is done by inserting the section into a loop of bacterial DNA (a plasmid). And there are many different reasons why one might want to go to the trouble of doing it.

Why engineer genomes?

Since research into genetic engineering began, the main focus has been in pure science. Studying the effects of "knocking out" certain genes can, for example, reveal huge amounts of new information about the way genomes and cells work. The first practical application of genetic engineering, in 1978, was the creation of a bacterium whose genome contains the human gene for making insulin. The transgenic bacteria produce insulin in large quantities in huge culture flasks. Before 1982, when the new medicine became widely available, patients with diabetes relied on insulin taken from pigs or other animals, which was more costly and more difficult, and some patients had allergic reactions to the nonhuman insulin. In 1996, molecular geneticists improved on nature by changing the gene sequence slightly; the resulting new strain of transgenic bacteria now make a slightly better insulin protein.

Many medicines are now made by transgenic bacteria (or sometimes transgenic yeast), and some commercial products contain proteins produced this way, too. For example, some low-fat, creamy icecream remains soft because it contains an antifreeze protein. The protein is made by transgenic yeast whose genomes contain the necessary gene, which came from the ocean pout (Macrozoarces americanus), a fish that lives in the Arctic Ocean.

Above left *Shoots of "golden rice," a genetically engineered strain of rice, busily making beta carotene, the precursor of vitamin A.*

Above right *Papaya fruit can now be genetically modified to make them resistant to papaya ringspot virus, one of the most destructive diseases to affect the crop, and it can only be partially controlled by conventional means.*

Going multicellular

A change in the genome of a single-celled organism, such as a bacterium, is passed down the generations. But to create a transgenic multicellular organism would mean changing the genome of cells that form the basis of the entire organism. When the organism matures, all of its cells will contain the transplanted gene. In the case of animals, scientists need to manipulate genes in cells of an embryo (see box). The process is easier in plants, because many plant tissues are capable of vegetative reproduction. Transgenic or "genetically modified" (GM) plants are often created to have resistance to disease or to herbicides and pesticides. It is not always that kind of quality, however, that is the aim of genetic modification of plants. The first genetically modified plant approved for human consumption was the Flavr Savr, a tomato engineered to have a longer shelf life.

Genetic modification of crops is controversial among some people, mostly because of vested interests of big businesses and because of fears surrounding the release of genetically modified organisms into the environment. However, some projects are less controversial than others. In the late 1990s, for example, the papaya industry of Hawaii was saved from a virus that was ravaging it after a gene taken from the virus itself was inserted into the papaya tree genome. In the first decade of the twenty-first century an international collaboration produced transgenic 'golden rice', which has two genes added to its genome: one from a daffodil and another from a bacterium. The new strain of rice produces a precursor to vitamin A, called beta carotene, in its grains. The transgenic rice is part of an effort to combat vitamin A deficiency, a major problem in some parts of the world.

TRANSGENIC GOATS

One application of genetic engineering involves transgenic goats, which manufacture a protein called antithrombin. The goats produce the protein in their milk; once purified, it is used in a medicine that prevents blood clotting during surgery and is also prescribed to people whose own bodies do not make the protein. To create the transgenic goats, the human gene was spliced to a section of goat DNA that causes genes to be expressed in the milk, and the resulting transgene injected directly into goat embryo cells.

In some of the embryos, the transgene was adopted as part of the genome. Those embryos were implanted into the womb of a female goat, and her female offspring expressed the gene, producing the protein in their milk. The descendants of those transgenic goats also have the transgene in their genomes, ensuring continued production.

Modifying the DNA
The gene that produces antithrombin is inserted into a section of goat DNA

Injecting the DNA
The modified DNA is injected into a fertilized goat egg

Implanting the DNA
The embryo is implanted into a female goat

Extracting the protein
The offspring create a herd of genetically modified goats

Duplicating—and creating anew

Two other well-publicized areas of scientific research involving manipulation of genetic material are cloning and synthetic biology. Cloning involves creating genetically identical organisms. In nature, any offspring produced by asexual reproduction is a clone, and so are identical twins. In the laboratory, cloning is normally carried out by nuclear transfer—removing the nucleus from one cell and inserting it into an enucleated egg cell, one that has had its own nucleus removed. The egg, with the new genome, can divide and produce a new multicellular organism.

Nuclear transfer was first carried out in the 1950s, when nuclei from the cells of frog embryos were implanted into enucleated frogs eggs. The nuclei from embryonic cells are primed, ready to kick-start the development of a new organism. However, it is much harder to do this when using nuclei taken from cells from an adult. The first somatic cell nuclear transfer—producing a clone of an adult organism using the nucleus of a somatic cell—was the well-publicized creation of a sheep called Dolly, born in 1996 (see box).

Synthetic biology is a bold and emerging field of study. It is an extension of genetic engineering, in which practitioners do not just manipulate existing genes but rather redesign them—or design completely

DOLLY THE SHEEP

The egg cell that would ultimately grow into Dolly was taken from an adult ewe. The egg was enucleated by physically removing the nucleus with a micropipette (a hair-thin device resembling a syringe). Next, a cell from a different ewe's mammary gland was "fused" with the enucleated cell using an electric pulse. The newly nucleated egg cell now contained a complete, different genome. Compounds were applied to the cell that would reprogram it so that it would behave like a fertilized egg despite the fact that it contained a nucleus from a mature, somatic cell. The pseudo-fertilized egg cell divided, forming an embryo that was implanted into yet another ewe, who later gave birth to Dolly. The process was very inefficient; the egg from which Dolly was produced was the only one of 277 eggs used in the experiment.

CELLS BEGET CELLS

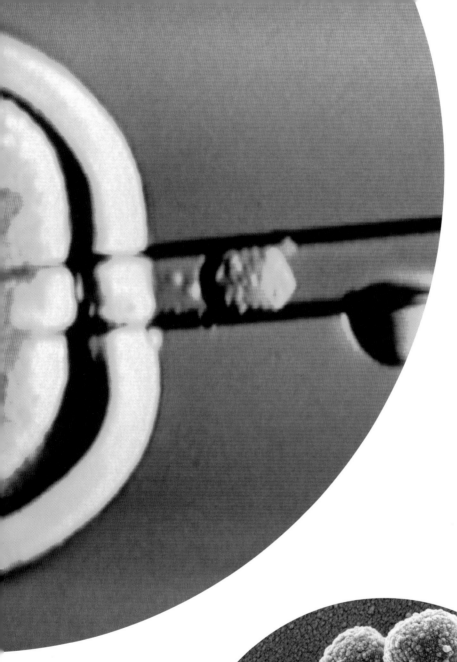

new ones. One of the aims of some synthetic biologists is to defy a rule that has stood since the beginning of life on Earth, the rule that states that all cells come from existing cells. Using DNA synthesizers, synthetic biologists have built genes, entire chromosomes, and even an entire genome, nucleotide by nucleotide. In 2010 a team of synthetic biologists managed to synthesize a complete genome of a simple bacterium called *Mycoplasma mycoides*, and then implanted it into an enucleated cell of a related, but different, species. The synthetic genome worked, and the cell was viable. (This is not yet a truly synthetic life, because only the genome was synthesized. All the other intricate molecular machinery inside the cell, and the membrane and cell wall, already existed.)

There are many potential benefits of synthetic biology. For example, organisms could be designed that capture the sun's energy more efficiently than ordinary photosynthetic organisms, offer promising new alternative energy sources or have the ability to clean up oil spills or digest plastics. The organisms chosen for such tasks would be bacteria—not only do they have far simpler genomes than eukaryotes but also, as single-celled organisms, they reproduce more quickly. These very qualities have helped to make bacteria the most adaptable and successful life form on the planet, the subject of the next chapter, where the focus is on bacteria and other single-celled organisms.

Above *Injection of a nucleus into an egg cell that has had its nucleus removed during the cloning procedure of a sheep. The incoming nucleus is derived from a somatic cell—a cell from ordinary tissues, such as skin.*

Left *Colored scanning electron micrograph of cells of* Mycoplasma mycoides *JCVI-syn1.0, the first self-replicating bacterium controlled by a synthetic genome. The bacterium's genome contains "watermark" sequences.*

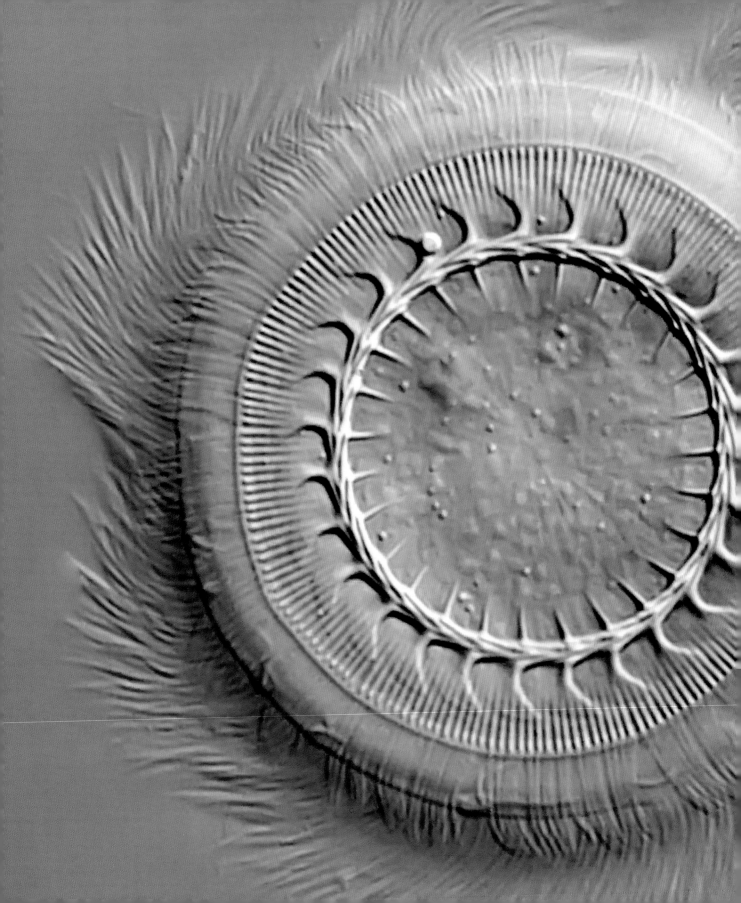

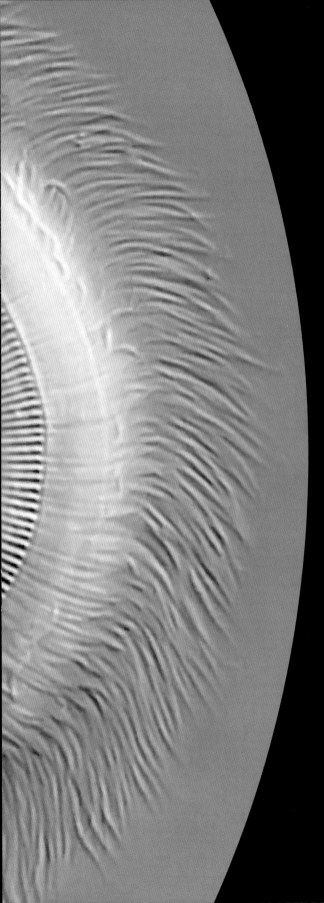

CHAPTER FOUR
Cellular Singletons

Everywhere we look, anywhere on the planet, every drop of water, the soil beneath our feet, even the air we breathe—all is teeming with living things too small to see, most of them existing as single cells. In fact, the great majority of cells on Earth are individual living things.

Left *Beautiful but deadly. Differential interference contrast micrograph of the single-celled eukaryotic parasite* Trichodina pediculus. *This organism lives on the gills and mucous membranes of many aquatic animals, where it stimulates the overproduction of mucus.*

Out there on their own

It can be hard to survive alone in a harsh world. All the functions necessary for life—respiration, digestion, excretion, locomotion, and reproduction—need to be carried out within a tiny fatty membrane. But unicellularity has its rewards, and single-celled organisms are opportunistic, diverse, and incredibly successful.

Success, everywhere

Existence as a tiny single cell might seem pretty limiting—especially for the majority of singletons, which are simple prokaryotic cells with an uncomplicated genome and no fancy organelles. But there are advantages to being small and simple and going it alone. Single-celled organisms can reproduce rapidly, and every mutation in every individual is in the germline (in other words, it will be passed on to subsequent generations)—so species can evolve much more rapidly. That is why single-celled life is so adaptable and exists in every corner of our world, including in very harsh environments. All of this adds up to rapid colonization of every habitat on Earth, even as those habitats change. So successful are single-celled organisms that they account for around 95 percent of all the biomass in the oceans and an appreciable proportion on land, too.

The diminutive size of single-celled organisms means that they can benefit from living inside other, larger organisms—including us. There are around ten times as many nonhuman cells inside our digestive tract than there are human cells in our entire body. Most of them are single-celled organisms, and nearly all are (prokaryotic) archaea and bacteria. Many benefit their human host, for example producing certain vitamins or helping to digest soluble fiber. Others are certainly not beneficial, leading to conditions such as cancer and inflammatory bowel disease. The good organisms normally keep the bad ones at bay. We are also colonized by microorganisms on the outside, feasting on the fats, sugars, and proteins of our dead skin; again, most of them are benign or beneficial. There are more single-celled organisms on our hands than there are human beings in the whole world, in hundreds of different species (see box).

Below *Colored scanning electron micrograph of a papilla on a human tongue with a small colony of bacteria.*

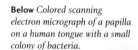

SINGLE-CELLED ORGANISMS AND US

A five-year initiative that ran from 2008 to 2013 identified more than ten thousand species of microorganisms that live in or on the human body. The great majority of them are single-celled, and the great majority of those are bacteria. The Human Microbiome Project, as this initiative was called, focused on areas of the body where micro-organisms are particularly rife—the digestive tract, the skin, the mouth, the vagina, the nasal passages, and the lungs—and aimed to establish the links between our single-celled cohabitees and our health. Mostly, the scientists involved carried out genomic studies of whole communities taken from these areas, inferring the many species present from their genomes-in-the-mix. They also conducted studies of the individual genomes of many of the organisms they found. One surprising finding of the project is that single-celled organisms contribute more genes necessary for our survival than we do ourselves.

It takes all kinds

Microorganisms can make the most of limited resources, and they positively thrive when times are good. They play incredibly important roles in the world's ecosystems—we couldn't live without our microscopic cousins. And yet, for all their success, the world's single-celled organisms are a ragtag bunch. The only thing they really all have in common is their unicellularity. Most live truly as individuals, although there are many unicellular species that come together and live as colonies (more of that in chapter five). Some are predators; some are parasites; some benefit from symbiosis (literally, meaning "living together") with other organisms, normally to the benefit of both parties, such as the good bacteria in our digestive tract.

There are single-celled organisms in all three domains of the natural world. In the domain Eukaryota, in which the best-known species are the multicellular animals, plants, and fungi, there are many species that exist as singletons. They include waterborne parasites and the oxygen-generating photosynthetic algae in the oceans. The most prolific inhabitants of the unicellular world, however, in terms of species and sheer numbers, are the Bacteria. Some bacteria are infamous because they cause disease but many more are beneficial, and while we could not live without them, they would be fine without us. The third domain, Archaea, is smaller than bacteria—there are a few hundred known species—but bigger and more important than we might imagine. Estimates suggest that Archaea accounts for about one-fifth of the world's biomass. And it is there that our tour of the unicellular world begins.

Left *Three singletons: a bacterium, an archaeon, and a eukaryote. 1. Colored transmission electron micrograph of the bacterium* Pseudomonas syringae *(see also page 100). 2. Colored scanning electron micrograph of* Methanosarcina mazei *archaea. 3. Light micrograph of* Euglena, *a eukaryote singleton.* Euglena *does not photosynthesize but it ingests small algae that do.*

CELLULAR SINGLETONS 93

Prokaryotic singletons

The majority of single-celled species belong to the domains Bacteria and Archaea; they are prokaryotic cells without a nucleus or organelles. While most people are more than aware of bacteria, many are not familiar with archaea—and nor was the scientific community until fairly recently.

Archaea—new kids on the block

For most of the previous century, biologists characterized all noneukaryotic organisms as bacteria and supposed they were all closely related to each other. But in the 1970s, a new kind of organism challenged that neat dichotomy. These were organisms, known since the 1960s, existing in extreme environments: hot springs and undersea hydrothermal vents where temperatures can reach boiling point; deep in soils at the bottom of the ocean where no light or oxygen is present; in lakes in which the concentration of salt would kill any life form previously known. One of these extremophiles, *Picrophilus torridus*, thrives in extremely concentrated sulfuric acid; another, *Methanopyrus kandleri*, happily reproduces at 250° Fahrenheit (122° Celsius).

The extremophiles caused a stir but biologists called them archaebacteria ("archae-" from the Greek word for "ancient"), because it seemed they might have been direct descendants of some of the earliest living things, back when conditions all over the planet would have been extreme. And despite the "bacteria" part of their name, studying archaebacterial genes revealed significant and consistent differences between them and bacteria. It was a long debate, but eventually in the 1990s, the Three Domain system was accepted into the mainstream. Since being accepted as a domain in their own right, archaeal organisms have been found in far less extreme locations, including ordinary soil and even human navels and on teeth.

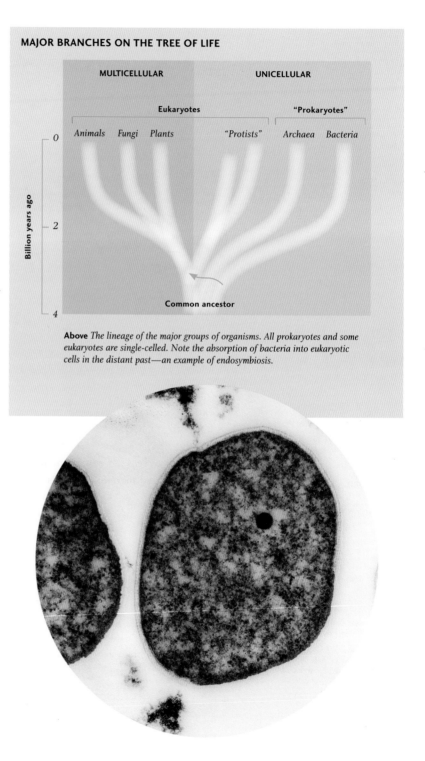

Above *The lineage of the major groups of organisms. All prokaryotes and some eukaryotes are single-celled. Note the absorption of bacteria into eukaryotic cells in the distant past—an example of endosymbiosis.*

Right *Grand Prismatic Spring, Yellowstone National Park, a 370-foot (112-meter) diameter body of water of which the temperature is a scorching 158°F (70°C). The vivid colors around the perimeter are due to pigments produced by thermophilic (heat-loving) archaea.*

Left *All archaea are single-celled organisms that are similar to bacteria but share some characteristics with eukaryote cells. This transmission electron micrograph shows a cross section through* Sulfolobus, *an archaeon that thrives in acidic and sulfur-rich environments and achieves optimal growth at 176°F (80°C).*

Planktonic (free-floating, individual) archaea in the ocean make up a large proportion of the ocean's biomass. Archaea have also been identified as the predominant organisms present deep in the layer of sediment covering the ocean floor. The archaea in this sediment are extremely slow living, consuming hardly any energy; they divide once every few hundred years but some parts of the sediment contain thousands of millions per cubic inch (hundreds of millions of cells per cubic centimeter). The best estimates suggest that archaeal cells locked in ocean sediments could constitute up to 10 percent of the entire biomass on Earth.

What are archaea?

Archaea have no nucleus or other organelles, therefore, they are prokaryotes like bacteria (see box on facing page) but the actual cellular machinery of transcription (making messenger RNA) and translation (building proteins) is more similar to eukaryotic cells. Archaean cells also have histone proteins—the beads on the "beads-on-a-string" arrangement of DNA found in eukaryote cells (see chapter two); bacteria do not have histones.

Genomic studies in the late 1990s showed that archaea are more closely related to humans than they are to bacteria. And yet, again like bacteria, archaea have a simple genome and are physically small. Their cell membranes also set them apart; they are made of slightly different fatty molecules from those of both bacteria and eukaryotes.

The domains of Archaea and Bacteria both existed before that of Eukaryota arose. The earliest known, deepest, branching on the tree of life (see chapter three) coincides with the last common ancestor of Archaea and Bacteria. This branching must have taken place very early in the history of life on Earth, more than three billion years ago. Eukaryotes almost certainly evolved from the archaeal branch instead of the bacterial one. In fact, eukaryotes probably originated from archaeal cells that engulfed but did not digest bacteria that were particularly good at generating energy. If this idea is correct, then the ingested bacteria were the ancestors of mitochondria and chloroplasts inside eukaryotic cells (this is the theory of endosymbiosis, mentioned in chapter two).

GALLERY OF EXTREMOPHILES

Nearly all extremophiles are archaea—but not all archaea are extremophiles. These examples are species that can survive in conditions of salinity, temperature, and pH that would be hostile to most organisms. Extremophiles are of interest to astrobiologists looking for life on other planets or moons in the solar system and beyond, where such extremes may dominate.

Halococcus archaea
Colored scanning electron micrograph showing clumps of Halococcus salifodinae, *a halophilic (salt-loving) archaeal species found in water with high concentrations of salt that would be deadly to most other forms of life.*

Methanosarcina archaea
Colored transmission electron micrograph of a section through several Methanosarcina rumin *archaea (cell walls are colored yellow).* M. rumin *is found in places with little or no oxygen, such as lake sediments, garbage dumps, and in the intestines of some mammals and insects. It produces methane gas.*

Thermophilic bacteria
Light micrograph of a sample of water from a geyser in Yellowstone National Park, containing various species of thermophilic bacteria that thrive in the high temperatures of the water. Also visible are the empty skeletons of diatoms, single-celled eukaryotes that are not able to survive the high temperatures.

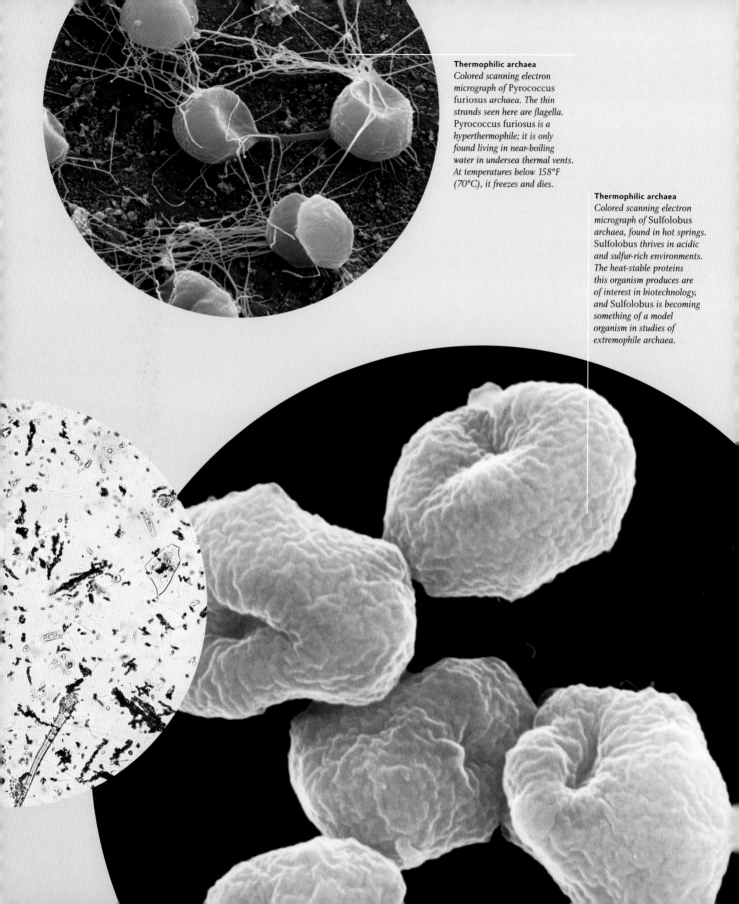

Thermophilic archaea
Colored scanning electron micrograph of Pyrococcus furiosus *archaea. The thin strands seen here are flagella.* Pyrococcus furiosus *is a hyperthermophile; it is only found living in near-boiling water in undersea thermal vents. At temperatures below 158°F (70°C), it freezes and dies.*

Thermophilic archaea
Colored scanning electron micrograph of Sulfolobus *archaea, found in hot springs.* Sulfolobus *thrives in acidic and sulfur-rich environments. The heat-stable proteins this organism produces are of interest in biotechnology, and* Sulfolobus *is becoming something of a model organism in studies of extremophile archaea.*

Classifying bacteria

And so to the largest domain, in terms of species and sheer numbers of individuals: Bacteria. There are thousands of bacteria in a drop of pond water or seawater, millions in a teaspoon of soil, trillions inside our colon. In each case, hundreds or thousands of species may be present. Bacteria come in different shapes—the main two being ball-shape (coccus) and rod-shape (bacillus). They also have differing tendencies in the way they cluster together. These characteristics are used—together with the use of gram staining, which indicates whether bacteria have a thick cell wall or not (see chapter two)—in efforts to identify them.

There is great diversity in bacteria; some are pathogenic (they cause disease), some benign; some are aerobic, others anaerobic; some are parasites, while others live in harmonious symbiosis with other organisms. Some are even predatory, hunting in packs. *Myxococcus xanthus*, for example, is a soil-dwelling species that forms swarms, secreting digestive enzymes that kill other bacteria. When other bacteria are scarce, these miniature predators even come together to form small fruiting bodies—some of the cells become spores that can survive until things improve (see chapter six).

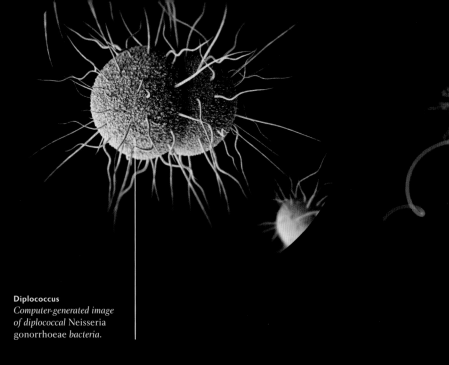

Diplococcus
Computer-generated image of diplococcal Neisseria gonorrhoeae *bacteria.*

Streptococcus
Computer-generated image of Streptococcus pneumoniae *bacteria. Streptococcal bacteria divide along only one axis.*

IDENTIFYING BACTERIA

These are the major morphologies (shapes, aggregation habits) of bacteria used to help identify different species.

- Coccus
- Bacillus
- Spiral bacterium
- Diplococcus
- Streptococcus
- Staphylococcus
- Diplobacilli
- Streptobacilli

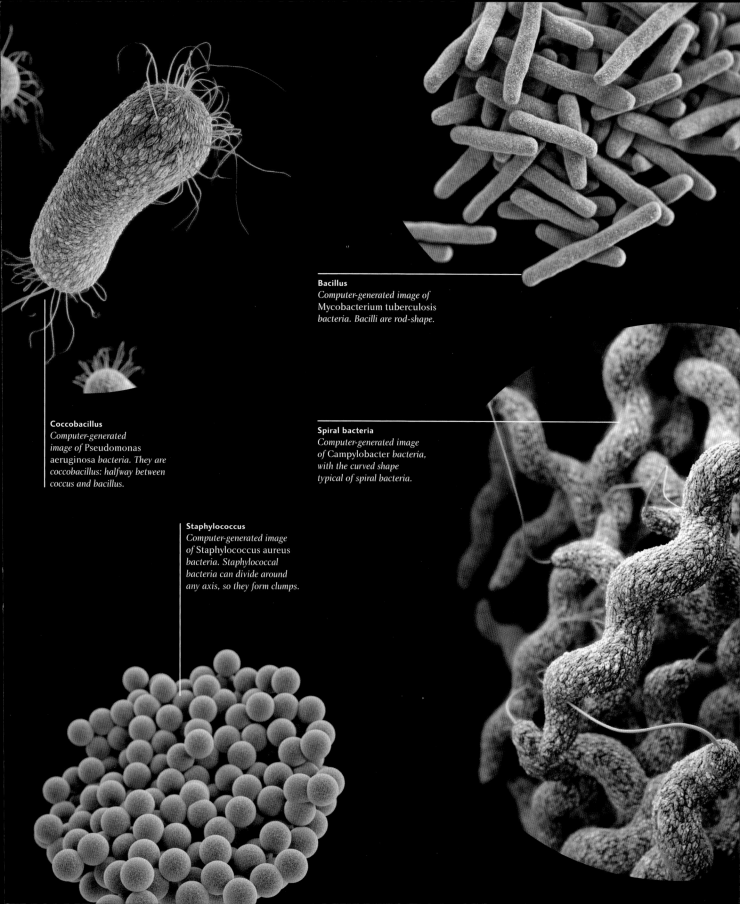

Bacillus
Computer-generated image of Mycobacterium tuberculosis *bacteria. Bacilli are rod-shape.*

Coccobacillus
Computer-generated image of Pseudomonas aeruginosa *bacteria. They are coccobacillus: halfway between coccus and bacillus.*

Spiral bacteria
Computer-generated image of Campylobacter *bacteria, with the curved shape typical of spiral bacteria.*

Staphylococcus
Computer-generated image of Staphylococcus aureus *bacteria. Staphylococcal bacteria can divide around any axis, so they form clumps.*

Above left *Snow-making machines use the bacterium* Pseudomonas syringae *to help ice crystals form.*

Center *Sandstone in Sedona, Arizona, which is colored red because of large amounts of iron oxide, the oxygen having been captured from the air by bacteria millions of years ago.*

Above *Termite, whose digestive system contains archaea that break down cellulose.*

Global influence

Species from Archaea and Bacteria play important roles in Earth's environmental cycles. One species of airborne bacteria, *Pseudomonas syringae*, is important in the water cycle. This bacterium produces proteins that mimic the shape of water molecules, and it attaches these proteins to its membrane so that water condenses there. This behavior encourages the production of rain and snow, affecting the weather globally. The ability of *P. syringae* to encourage water vapor to condense evolved as a way of enabling this bacterium to infect plants. When landing on plants' leaves in cold but not freezing weather, the bacterium causes ice to form. The ice fractures the plant cells, letting the bacteria into the plant. Since the 1970s, *P. syringae* have been put to use in snow-making machines in Alpine resorts.

The biggest global influence that prokaryotes have is in the carbon cycle; most produce carbon dioxide but many also fix atmospheric carbon, binding carbon dioxide into carbohydrates by photosynthesis. One phylum (large grouping) known as cyanobacteria (sometimes referred to as blue-green algae, although they are not algae) carry out photosynthesis on a massive scale. Cyanobacteria float around in all of the world's oceans, lakes, and rivers, producing oxygen whenever light falls on them. The direct but very distant ancestors of cyanobacteria were the very first organisms to produce oxygen, around 3.5 billion years ago. At that time, Earth's atmosphere contained virtually no oxygen; it was a reducing atmosphere (see page 54), rich in hydrogen and carbon monoxide. As cyanobacteria began to thrive, the oxygen they produced dissolved in water, oxidizing dissolved iron to form rusty deposits. As this process ebbed and flowed, the concentration of iron-oxide sediment rose and fell, producing sedimentary rocks called banded iron formations.

Eventually oxygen made it out of the seas and into the atmosphere, which has remained oxygenated ever since, and the chloroplasts inside modern-day algae, probably descendants of ancient cyanobacteria engulfed by early eukaryotic cells (see page 55), continue to pump out oxygen today. Some archaeal cells play a different role in the carbon cycle, producing methane (CH_4), the only organisms that do so. These methanogens are present in marshes, in sewage sludge, and in the digestive flora of humans, cows, and termites, for example, where they help break down cellulose.

Grabbing nitrogen

Many species in Bacteria and Archaea also play a vital role in the nitrogen cycle. Nitrogen makes up four-fifths of the atmosphere, but as molecules made of two nitrogen atoms bound tightly together (N_2). Nitrogen is essential for making protein molecules and nucleic acids, so it is literally vital that those bonds are broken and the atoms fixed into molecules that living things can use. Nitrogen-fixing prokaryotes take in nitrogen molecules and produce compounds that contain ammonium (NH_4). Some atmospheric nitrogen is also fixed by lightning—it produces nitric oxide (NO)—but nitrifying archaea and bacteria fix around 210 million tons (190 million tonnes) per year, more than twenty times as much as lightning does. Without the nitrogen-fixing archaea and bacteria, Earth's ability to support diverse forms of life would have been severely limited.

Bacteria from the genus *Trichodesmium* fix more nitrogen in the world's oceans than any other. Trichodesmia are filamentous bacteria; as they reproduce, the cells do not divide properly, so they remain together as a very, very long single cell containing many copies of its genome. These remarkable bacteria can sometimes be seen as large blooms and have earned the nickname "sea straw."

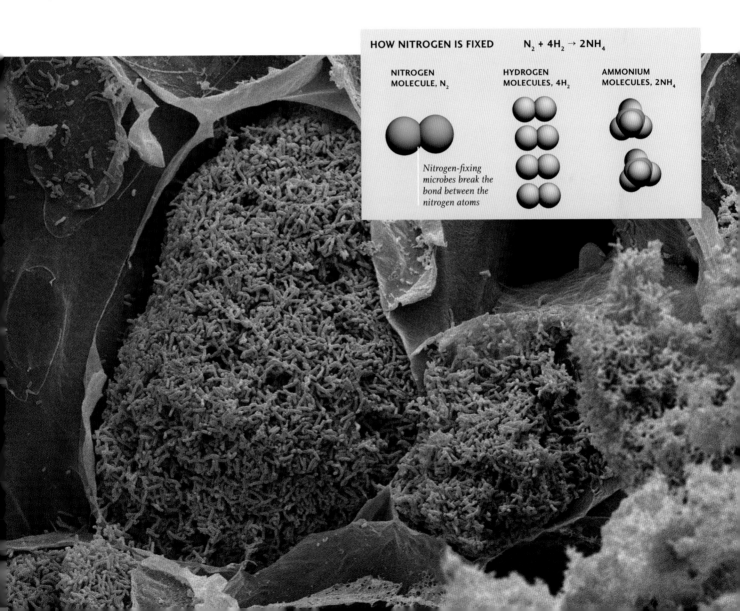

Below *Colored scanning electron micrograph of* Rhizobium leguminosarum *nitrogen-fixing bacteria (brown) inside ruptured root nodule cells of a runner bean plant (*Phaseolus coccineus*).*

HOW NITROGEN IS FIXED $N_2 + 4H_2 \rightarrow 2NH_4$

NITROGEN MOLECULE, N_2

HYDROGEN MOLECULES, $4H_2$

AMMONIUM MOLECULES, $2NH_4$

Nitrogen-fixing microbes break the bond between the nitrogen atoms

Working together

Nitrifying bacteria are also found in soil, and the nitrogen-rich compounds they produce dissolve in groundwater, which is taken up by plant roots. Some form mutually beneficial symbiotic relationships with certain plants—in particular pulses, such as clover. Many bacterial species, collectively referred to as *Rhizobia*, only fix nitrogen when they are associated with a plant. Receptors in the cells of a plant's roots sense nitrifying bacteria nearby; the gene expression in the roots changes, and a protective, nurturing nodule grows around the bacteria. The plant supplies starch as food for the bacteria, and the bacteria release nitrogen-rich compounds that the plant needs.

There are many other examples of bacterial symbiosis—beneficial, in most cases, to both the bacteria and the other organisms. Luminescent (light-producing) bacteria live inside certain organs of marine animals. Once again, they benefit from a safe environment and from being supplied with nutrients—and the light they produce, on demand, is very useful to the organisms in which they live. In the deep-sea anglerfish, for example, bacteria live in the small dangly organ (called an esca), which the anglerfish uses to attract small prey.

Another organism to evolve a close relationship with luminescent bacteria is a squid called *Euprymna scolopes*. All squid are masters of camouflage—they can change the color and texture of their mantle. *Euprymna scolopes* can even get rid of its own shadow on moonlit nights, thanks to a remarkable symbiosis with the bacterium *Aliivibrio fischeri*. These squid have light-sensitive cells on their upper (dorsal) surface, and colonies of *A. fischeri* bacteria live inside organs on the underside (ventrum). The squid signals to the bacteria, which produce just enough light to make the shadow disappear.

Bacterial symbioses also occur around hydrothermal vents, where undersea volcanoes heat water. The hot water carries various dissolved gases, such as methane, hydrogen, and hydrogen sulfide, that are useful but normally inaccessible or toxic to most living things. Bacteria living in the gills of bivalves (two-shelled mollusks) are able to access the nutrients, thanks to their unusual metabolisms and to supply them to their hosts in return for a safe place to live.

Another remarkable bacterial symbiosis involves ants. Many species of ant farm fungi, but their crops are under constant threat from a different parasitic

Below left *Nitrogen-fixing nodules on pea roots (Pisum sativum).*

Center *Hydrogen sulfide oxidizing bacteria in the gills of the bivalve* Pegophysema philippiana. *The bacteria are false colored.*

Below *Deep-sea angler fish with its esca illuminated by luminescent bacteria.*

CELLULAR SINGLETONS

fungus. The ants grow cultures of bacteria from the genus *Pseudonocardia* on their cuticle (outer covering). These bacteria produce a chemical compound that restricts the growth of the parasitic fungus—and everyone is happy (apart from the parasite).

Bacteria and disease

Archaea and bacteria do some wonderful things: they make oxygen, fix nitrogen, and live in harmony with loads of other types of organism. However, they are not all sweetness and light. Many of them infect animals and plants, causing life-threatening diseases. (Actually, that is not fair on the archaea, which so far are not implicated in any diseases. And there are plenty of eukaryotic single-celled organisms that are pathogens, too, as we will see.)

Throughout human history, bacterial infections have resulted in hundreds of millions of early deaths. In some cases, bacterial infection is opportunistic: a cut on a finger provides access to the warm, nutritious internal environment of a person's body. But it is not always opportunistic; some pathogenic bacteria have evolved cunning ways to pass from one host to another, greatly enhancing their chance to thrive and reproduce. Shown here and on the following two pages are some of the most common human pathogenic bacteria that cause widespread disease, suffering, and death.

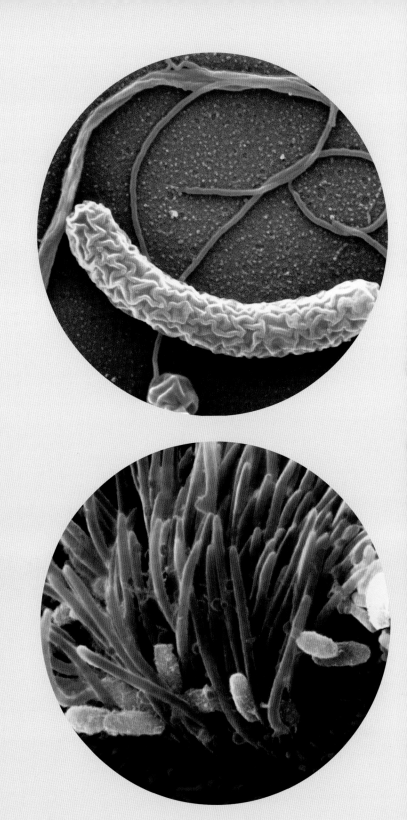

Top right Vibrio cholerae, which causes cholera, thrives inside the intestines. Its progeny are expelled into watercourses to be ingested by other animals.

Right Bordetella pertussis (yellow), which causes whooping cough, lives on ciliated cells in the respiratory system. It is either sneezed or coughed out into the air to be breathed in by other animals.

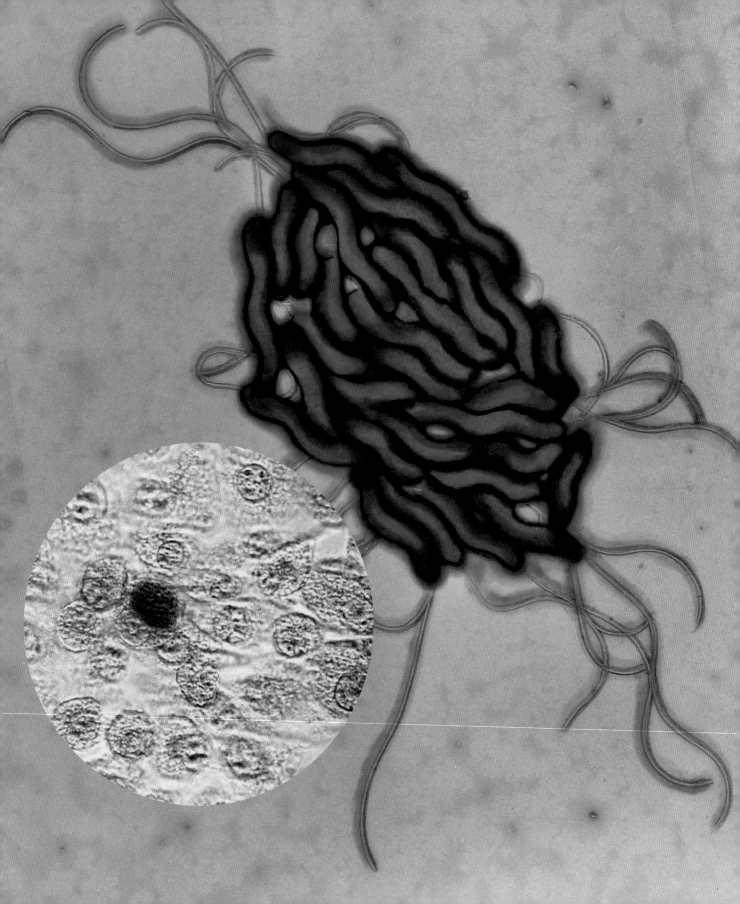

Left Campylobacter jejuni, which is the main cause of food poisoning worldwide, can live in any plant or animal. It is simply passed from one to another when an animal eats an infected plant or animal, and it can also be carried by flies.

Left (inset) Chlamydia trachomatis, which causes chlamydia, lives inside cells in human reproductive and urinary systems. It can be passed from person to person by close sexual contact, and from mother to baby during childbirth.

Right Mycobacterium tuberculosis, which causes tuberculosis, reproduces inside cells in the lungs. It is spread by being sneezed out and breathed in.

Below Yersinia pestis (colored yellow), which causes the bubonic plague, on the ridges of the oriental rat flea (Xenopsylla cheopis). It can happily live in either the bloodsucker or the bloodsuckee.

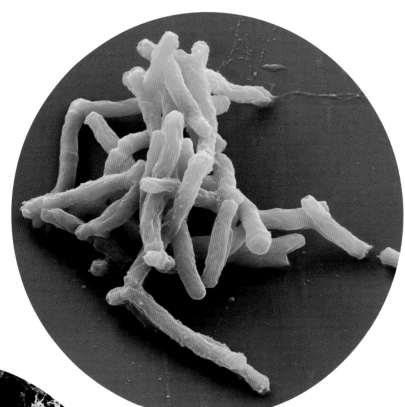

Many pathogenic bacteria have evolved to be fairly benign; the illness they cause will not kill. This is normally the case when transmission of the bacteria depends upon the ability of the victim to move around or be active. In cases where that is not important—for example, in waterborne diseases such as cholera—the bacteria has not evolved any such benignity. Even in those cases, however, it is not generally worth the bacterium's while killing the victim—it is normally the body's efforts to get rid of the bacteria that actually kills. In cholera, for example, the body leaks water into the intestines, trying to flush the bacteria away; the result is severe dehydration and extreme diarrhea. In the case of the bubonic plague, the bacteria live in the lymph nodes and migrate to the lungs, but it is a whole-body (systemic) reaction to the infection that can kill the human victim: a septic shock.

Similarly, the uncomfortable symptoms of nonlethal infections are the result of the body's efforts to limit the proliferation of the bacteria. The high fever that results from bacterial infection is part of that effort, because pathogenic bacteria reproduce best at normal body temperature. And the streaming nose is the body's attempt to prevent bacteria (or viruses) from reaching the lungs. The pus that fills infected boils is made of dead white blood cells, the body's defense team. The redness and throbbing pain are the result of increased blood flow to the infected area, whose aim is to bring increased amounts of various blood products that help to fight infection and heal the damaged tissue.

One of the most famous pathogenic bacteria in recent years is "MRSA": methicillin-resistant *Staphylococcus aureus*. This bacterium is a particular strain of *S. aureus*, a species that lives naturally on the human body; the strain that hits the headlines is resistant to an antibiotic called methicillin. In fact, it is resistant to a whole class of antibiotics. This strain of bacteria has acquired its resistance through natural selection; once the ability to resist the antibiotic had arisen as a random mutation, those bacteria that inherit that trait would have a much better chance of surviving and would reproduce, enabling the mutation to gain a foothold in the bacterial gene pool. MRSA infections—and bacterial resistance to antibiotics in general—have become a major issue in public health.

Model organism
Another bacterial species that sometimes hits the news, and often presents a headache to public-health officials, is *Escherichia coli*. In fact, most strains of *E. coli* are perfectly harmless to humans—not only harmless but beneficial. *Escherichia coli* bacteria are among the good prokaryotes inside the colon; they produce vitamin K and help to keep pathogenic bacteria at bay. However, there are many virulent strains, and if they find their way out of the colon, they can cause disease. If pathogenic *E. coli* make it to the peritoneum (the wall of the abdomen)—perhaps through a perforation in the colon—they can cause a nasty bout of peritonitis. Likewise, if certain strains find their way into the urinary system they can cause painful urinary tract infections. If they are swallowed by someone eating food infected with pathogenic strains of *E. coli* (the unpleasantly named fecal-oral route) they can cause gastroenteritis, which is accompanied by severe vomiting and diarrhea. It is through such outbreaks of food poisoning that most people become aware of this little bacterium.

Despite its bad press, *E. coli* is more useful than it is harmful. It is used as a model organism in microbiological laboratories across the world—and studies of its behavior, biochemistry, and genome have made *E. coli* the most intimately known of all the prokaryotes in the living world. Easy to culture and quick to reproduce, it has been one of the most commonly used bacteria in genetics experiments since the late 1940s. The importance of *E. coli* grew significantly in the 1970s, after it became the focus of early experiments in genetic engineering—and it is still the organism of choice for genetic engineers, for whom it dutifully makes copies of genes from other organisms or manufactures the proteins for which those genes code (see chapter three).

The genome of *E. coli* is 4.6 million base pairs long and contains nearly 4,300 genes—and, as is true of all prokaryotic organisms, it exists as a single loop of DNA. It was one of the first genomes to be sequenced, in 1997. Remarkably, only about 20 percent of the DNA is shared among the various strains of *E. coli*, which means there is up to 80 percent variation within this single species. In a project entitled "The Long Term Evolution Experiment" at Michigan State University, experimenters have been watching that variation shift over thousands of generations since 1988. Colonies of *E. coli* evolved significant traits not present in the founding species over 30,000 generations: genuine evolution in less than 20 years.

Right *Accurate artistic representation of a cross section through part of an* Escherichia coli *bacterium, by microbiologist and artist David S. Goodsell.*

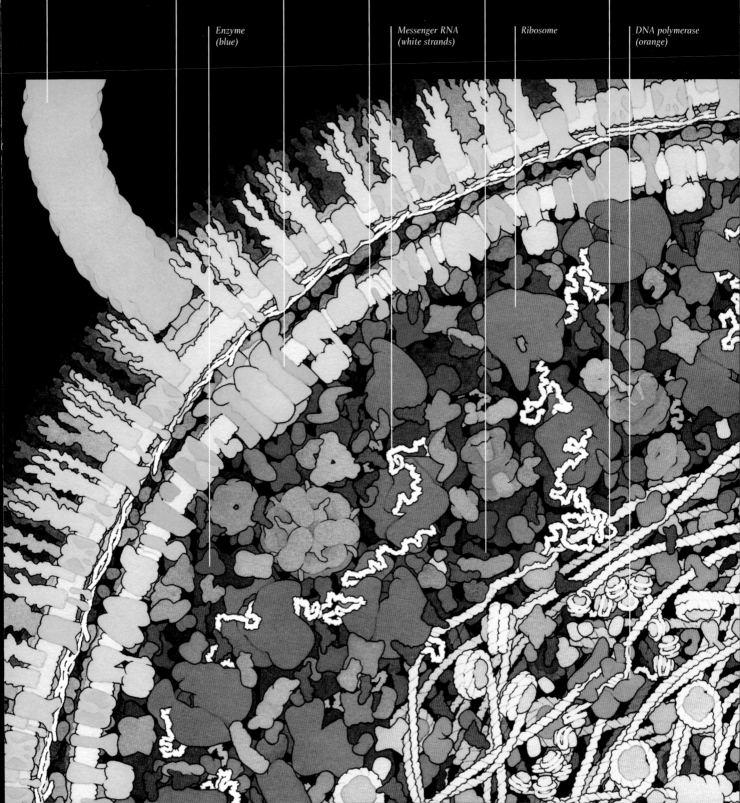
Enzyme (blue)   *Messenger RNA (white strands)*   *Ribosome*   *DNA polymerase (orange)*

Eukaryotic singles—protists

As one might expect, single-celled eukaryotes are generally much larger and more complex than their prokaryotic counterparts. The generic name for these organisms is protists. The variety of protists is huge, including some mind-bendingly beautiful ones as well as others that are somewhat deadly.

Classifying protists

Making sense of the world of single-celled organisms has long been considered a nightmare, especially when it comes to the eukaryotes. Genetic studies are helping to sort out which tiny eukaryotic organism is related to which other tiny eukaryotic organism but the new knowledge gained in the genomic age is having to battle hard against long-standing and commonsense—but ultimately flawed—systems of classification.

Today, the word "protist" is a convenient catch-all term for unicellular eukaryotes, but when German biologist Ernst Haeckel came up with the term, in 1866, it included what we now know as prokaryotes. At that time, when biological classification was in its infancy, everything was considered plant or animal. Single-celled organisms were not seen as an exception: animal-like single-celled things were called protozoa, while plantlike ones were protophyta. Haeckel and his contemporaries supposed that protists were closely related to each other by virtue of the fact that they are all single-celled. Haeckel's Protista kingdom even included bacteria. But bacteria, and eventually archaea, broke away from this classification, and ever since, biologists have been trying to rework the tree of life.

Comparing genomes (or often, more practically, just a selection of genes) between species really helps. But the bewildering array of different single-celled eukaryotic species makes the task incredibly difficult, and the result challenging to comprehend. There are

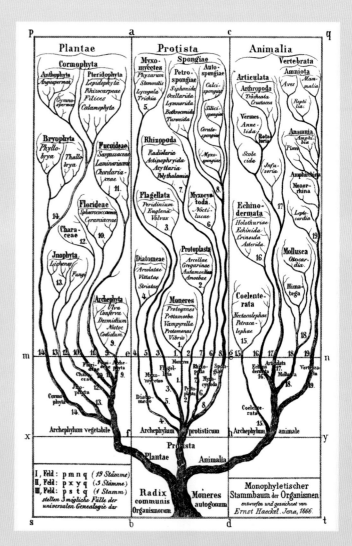

Above *Ernst Haeckel's tree of life, 1866, based on Charles Darwin's theory of evolution. Notice the three kingdoms: Plants, Protists, and Animals.*

protists in several separate, unrelated branches of the tree of life. The whole thing is confusing and unresolved. Yet all of this takes nothing away from the sheer beauty and fascination of some single-celled eukaryotic organisms. The best way to think about it is simply not to worry. For the purposes of this chapter, we will treat single-celled eukaryotes as being plantlike, animal-like, or funguslike—something with which Haeckel and his contemporaries would have felt entirely comfortable. Just remember that each of these categories includes organisms from all the major, unrelated branches of Eukaryota.

Variety among the protists

Unsurprisingly for such an unrelated collection of organisms, protists come in a vast array of shapes and sizes. They also have a wide variety of lifestyles, with a bewildering array of different life cycles occurring among the single-celled eukaryotes. All are capable of sexual reproduction, but many reproduce asexually most of the time. The funguslike and plantlike protists reproduce through the alternation of generations, as described in chapter three. There is variety, too—although perhaps less so—in the way protists obtain their energy. Many survive by ingesting other cells; many photosynthesize (some using organelles slightly different from chloroplasts, called plastids). Yet others simply absorb nutrients that are dissolved in their surroundings. The great majority of protists respire aerobically when they can but some have no need of oxygen and can only survive in low-oxygen conditions.

The smallest known protists are the picophytoplankton. That prefix, "pico-," means "very small" and "'phyto-" means "plant"—so picophytoplankton are very small free-floating plantlike things. The smallest of all is *Ostreococcus tauri*. It is just 1 micron or 0.001 millimeter in diameter, so small it appears as a tiny dot in an optical microscope. It is typically only big enough to squeeze in one mitochondrion and one chloroplast. The largest-known protist is *Syringammina fragilissima*, which grows as a sphere and can be over 8 inches (20 centimeters) in diameter. It achieves this by a process of cell division that does not actually involve the division of the parent cell. This protist is an undivided multinucleated mass of many generations and contains hundreds or thousands of nuclei.

Cell size is normally limited by the need to have a surface area large enough to serve its volume. The smaller the surface area, the fewer essential gases and nutrients and wastes can diffuse across the membrane. But, counterintuitively perhaps, making a cell bigger

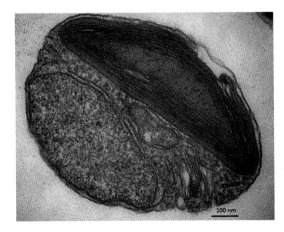

Right *Transmission electron micrograph of* Ostreococcus, *the smallest-known protist, which measures less than 1 micron across—smaller than most bacteria.*

Far right Syringammina, *the largest-known protist, grows to over 8 in (20 cm) in diameter.*

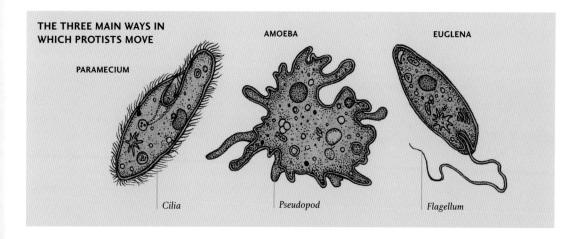

THE THREE MAIN WAYS IN WHICH PROTISTS MOVE

PARAMECIUM — Cilia
AMOEBA — Pseudopod
EUGLENA — Flagellum

makes the situation worse, not better. Double the diameter and the surface area quadruples—but the volume is now eight times what it was. This is why cells are generally very small. *Syringammina fragilissima* overcomes the problem by having deep lobes in its surface, effectively increasing the surface area many times. The lobes are so deep that they are in effect tubes extending into the cell's body. They are supported by a rigid outer casing, which the organism constructs by gluing together the tiny shells of other protists.

Syringammina fragilissima minces along the ocean floor engulfing detritus as it goes. Its movement is amoeboid, meaning that it takes advantage of its exclusively eukaryotic, dynamic cytoskeleton to change its shape and squirm along. Amoeboid movement is common in many protists. Many others live a planktonic existence, drifting in oceans, rivers, and lakes, many using flagella or cilia to help them get around.

Protist movement

A large number of protists move around by virtue of the coordinated beating of hundreds of cilia. These ciliated protists are among the most complicated and beautiful single-celled organisms there are. Sensibly enough, they are called ciliates—and, refreshingly enough, all are fairly closely related. The ciliates are as efficiently organized as they are complicated. Each one has two nuclei: the smaller one is a germline nucleus, which remains untouched apart from when it is time to reproduce; the other one is a somatic nucleus, for everyday tasks of gene expression that build enzymes and other proteins.

Although many ciliates appear similar to algae—translucent and green—there are no ciliates that photosynthesize. They are all predators, feeding on other tiny organisms—prokaryotes and protists. (When they appear green, it is because they have ingested green algae.) They use their everrestless cilia to sweep prey toward a mouthlike indentation called a cytosome. The cytosome absorbs the prey, wrapping it into a membrane-bound vacuole for digestion inside the cell. One of the most common ciliates, *Paramecium*, can swallow up to five thousand victims a day. With a "mouth", a "stomach," and a voracious appetite, it is easy to assume that these tiny creatures are like animals—and they are indeed among the animal-like protists that Haeckel called protozoa (the suffix "-zoa" means "animal").

Other examples of what Haeckel called protozoa are the dinoflagellates. Their name, rather charmingly, means "possessors of whirling whips"; as it suggests, these organisms get around using flagella instead of cilia. There are other flagellated protists, including plantlike and even funguslike ones (although only in some parts of their life cycle).

Right Macro photograph of Noctiluca scintillans, *a spherical, flagellated protist that can grow up to $1/16$ in (2 mm) in diameter, commonly found in marine plankton. It is also called "sea sparkle," because it is bioluminescent and large numbers give the ocean a gentle blue glow.*

Below Phase contrast light micrograph showing a group of Vorticella, *a ciliate protozoan (animal-like protist). Each single cell has a long stalk that anchors it to a surface.*

Below right Colored scanning electron micrograph of Dendrocometes paradoxus, *a protozoan that lives on the gills of freshwater fish and feeds by catching passing particles in its tentacles.*

CELLULAR SINGLETONS

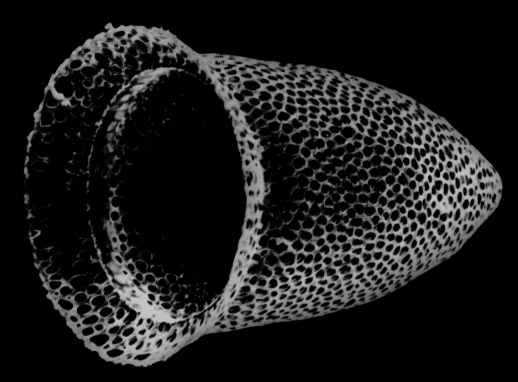

Left *Colored scanning electron micrograph of an empty lorica (shell) of a (single-celled) marine ciliate protist in the order* Tintinnida. *The lorica is made of an organic substance known as tectin, which the organism secretes, often strengthened by absorbed minerals.*

Below *Colored scanning electron micrograph of the diatom* Campylodiscus costatus, *with its protective frustrule (mineralized cell wall) made of silicon dioxide. Diatoms are commonplace in the oceans, and their frustrules form a clay sediment called diatomaceous earth.*

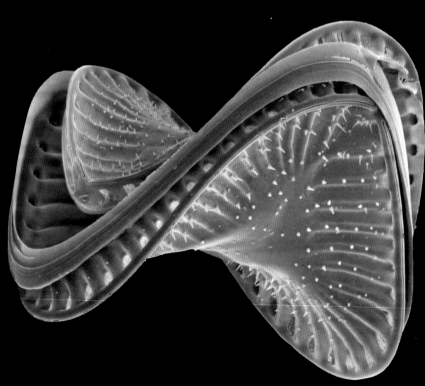

Right *A chalk cliff, part of England's White Cliffs of Dover. Chalk is a sedimentary rock made of the coccoliths (mineral outer casings) of marine protists, mostly coccolithophores, such as Gephyrocapsa oceanica, pictured below.*

Hard cases

Many protists that live in water grow intricate protective outer casings. Different groups use different materials, and the structures have various names. Some ciliates and flagellates build loricae, basket-shape cases made of a variety of different biological materials, depending upon the species. The material to build a lorica is produced constantly inside the cell and held inside tiny granules, which can be seen through a microscope.

Other organisms called diatoms build frustrules, which they make out of dissolved silicate minerals. The resulting structures fall to the bottom of the lake or ocean where the organisms live, forming a soft, crumbly, absorbent sediment known as diatomaceous earth. It has many uses—gardeners prize it as an insecticide, for example. In the 1860s, diatomaceous earth's absorptive quality was the solution to a problem of how to transport nitroglycerin, a very vibration-sensitive liquid, without it exploding at the wrong time. The result could have been called diatomaceous earth-soaked nitroglycerin, but it was given a rather more memorable name: dynamite.

Plantlike flagellates called coccolithophores build coccoliths, made from calcium carbonate. Amoeboid protists called foraminifera also produce hard cases made of calcium carbonate—known as tests—but here the compound has a slightly different crystal structure. The coccolithophores are "chalk" while foraminifera deposits are "limestone," but both forms of calcium carbonate occur in huge rock formations all over the world. The famous White Cliffs of Dover, on the south coast of England, is formed of an enormous mass of countless coccoliths, lifted up out of the ocean by tectonic forces.

Dangerous protists

Just as there are pathogenic bacteria, many protists can also cause diseases. The most infamous are members of the genus *Plasmodium*, for these are the organisms that cause malaria, a disease that claimed two million lives a year in the first half of the twentieth century and has probably caused more human deaths than any other infectious disease. The life cycle of *Plasmodium* is a complex drama that plays out at microscopic scale inside three hosts: one mosquito and two humans. The mosquito does not suffer as a result but the drama can easily turn into tragedy for the two humans.

The first human is someone who is already acting as host to *Plasmodium*. At this stage, the *Plasmodium* organisms live inside red blood cells in that person's body. They reproduce ferociously, bursting out of their cellular hosts to infect other red blood cells. Some of the individuals undergo meiosis, producing gametes, and some of these gametes are taken up through the invading proboscis of a hungry bloodsucking mosquito. Inside their new host, the gametes produce a zygote, which reproduces to form spores. These spores are passed to a new human host whom the mosquito settles on for its next meal. The spores infect cells in the liver of the new host, and they change form again, burst out into the bloodstream, and infect red blood cells, so beginning the whole process once more. The plasmodium has successfully sustained its species, the mosquito has had two meals—but the two humans, apart from itching from the mosquito's bite, are now carrying a life-threatening parasite.

A malarial infection causes a reduction in blood circulation, because when the plasmodium is inside red blood cells, it causes them to stick to blood vessels. This seems to have evolved as a way of ensuring the infected blood cells don't pass through the blood-filtering, parasite-killing spleen. The reduction in blood flow lowers the amount of oxygen supplied to vital organs. In addition, the sheer mass destruction of infected cells may cause severe anemia (reduction in the number of oxygen-carrying red blood cells).

Left *A mosquito from the species* Anopheles stephensi, *which has punctured a person's skin and is sucking blood. These insects can act as a vector (carrier) of* Plasmodium, *the protists that cause malaria.*

Right *Light micrograph of red blood cells, one of which is infected with* Plasmodium vivax, *one of two species of* plasmodium *that produces characteristic dark spots called Schüffner's dots.*

Below *Scanning electron micrograph of* Giardia *protists on the surface of the small intestine of a gerbil.*

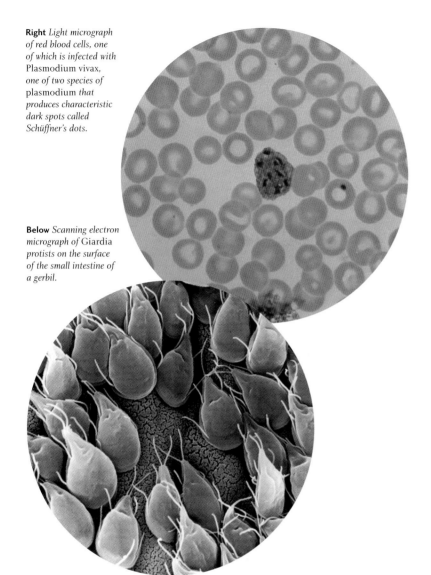

Plasmodium is a member of a phylum of protists called Apicomplexa. All members of Apicomplexa are protozoa (animal-like protists). All Apicomplexa spend part of their life cycle inside the cells of animal hosts, just as plasmodium does. These organisms have an organelle called an apicoplast, which is a kind of spear with which the organisms can break into other cells. Other Apicomplexa include *Toxoplasma gondii*, the cause of toxoplasmosis, and *Cryptosporidium parvum*, the cause of cryptosporidiosis.

Plasmodium is all too widespread in the tropics, but is not really a problem at more temperate latitudes. The most prevalent disease-causing parasite in humans worldwide is another protozoan, called *Giardia lamblia*. This tiny, flagellated organism is responsible for a common disease called giardiasis—with symptoms of severe vomiting, diarrhea, bloating, and flatulence.

There are also many pathogenic, parasitic protists that cause diseases in plants. The most infamous in this category is *Phytophthora infestans*, which causes potato blight. This tiny pest can devastate potato crops on a massive scale, and it was responsible for three major famines in the mid-nineteenth century, including the 1845 Irish Potato Famine. Once again, such a huge effect for such a tiny organism.

Right *Leaves of a potato plant (*Solanum tuberosum*) damaged by the potato blight fungus (*Phytophthora infestans*).*

CELLULAR SINGLETONS

Much room for doubt

Phytophthora infestans is classed as an oomycete (which means "egg fungus"), a type of organism still sometimes referred to by its former name: water mold. Other pathogenic species of water mold exist, including those in the genus *Saprolegnia*, which live on fish scales and cause a life-threatening disease called fish fungus. This disease can spread rapidly in stagnant water, and the organism can continue to live on dead fish, releasing spores that can infect living fish. The name water mold is a nod firmly in the direction of fungi. *Phytophthora infestans* and *Saprolegnia* are funguslike protists, because they produce spores, just as fungi do, and they absorb food from the soil or water where they live, opportunistically scavenging from either living or dead organisms.

There is another kind of organism that is considered as a funguslike protist: slime molds. These, too, produce spores and, therefore, have a passing resemblance to fungi. Indeed, Haeckel (back in the late nineteenth century) classified water molds and slime molds as fungi, which he placed in the Plant kingdom. So, although he was the one who came up with the kingdom Protista, he put these tiny things in with plants. In keeping with the awkward, unexpected unrelatedness among the protists, the slime molds are not closely related to the water molds. Most slime molds are not even related to each other; they are found across several different branches of the tree of life. Furthermore, neither water molds nor slime molds are actually fungi; they do not have cell walls made of chitin, as fungal cells do. Genuine fungi that exist as single cells are called yeasts—which, despite their eukaryotic, unicellular natures, are not considered as protists. And, just to confuse things a little more, genuine multicellular fungi, which exist as networks of fibers called hyphae, are called … molds.

Part of the reason why water molds and slime molds are considered funguslike is that both these types of organisms appear as multicellular organisms for at least part of their life cycle. The most remarkable in this regard are the slime molds.

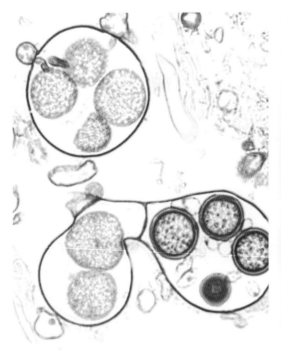

Far left *Light micrograph of the freshwater mold* Saprolegnia, *a protist often wrongly described as a fungus.* Saprolegnia *can feed on the scales and waste products of fish, causing an infection known as cotton mold.*

Left *A lithographic print of various marine protists, from Ernst Haeckel's* Kunstformen der Natur *(Artforms of Nature), a series of remarkable drawings of plants, animals, and protists first published between 1899 and 1904.*

Above left *The plasmodial slime mold* Fuligo septica, *commonly known as the dog vomit slime mold, moving slowly across a forest floor.*

Above right *The cellular slime mold* Stemonitis axifera—*a collection of many (single-celled) protists coming together to form erect spores on a wooden surface.*

There are two main types of slime mold: plasmodial and cellular. The plasmodial slime molds form a plasmodium (confusingly, because this has nothing to do with *Plasmodium*, the protozoan that causes malaria). A plasmodium is a single, amoeboid cell that slimes its way across dead or decaying plants, growing ever larger as it ingests food. This flat, expanded bag of cytoplasm can reach 10 inches (25 centimeters) in diameter. Inside, the cytoskeleton is working hard to push and pull the cytoplasm around to make the whole thing move—up to a yard (about a meter) a day. Nuclei are being duplicated all the time (by mitosis) but the cell remains as one (multinucleated) mass. Often, the plasmodium is brightly colored and, well, slimy.

When conditions are just right, an extraordinary thing happens. The plasmodium stops growing, stops moving and self-organizes into fruiting bodies. Inside a fruiting body, haploid spores form by meiosis. These can fuse with other spores from slime molds nearby and form a zygote, which will go on to become a new plasmodium.

More extraordinary, perhaps, are the cellular slime molds. They spend most of their lives as individual single cells—amoeboid protists. When food is scarce, they produce a chemical signal, and when the concentration of that chemical signal is high enough, the individual cells move together and form a slug-like mass called a grex. This multicellular "thing" now slimes around, its cells perfectly coordinated. And if we were to slice one of these things up, the cells can find their way back together and form a grex again.

Eventually, the grex will change, forming spores. Some of the cells form a sturdy base, some a thin stalk, and, at the top, some form the spore head. Only the cells in the spore head will divide by meiosis and have a chance of producing progeny for the next generation. This one-for-all, all-for-one approach is normal in a multicellular organism—but the cellular slime mold spore is made up of cells that have existed as completely individual units. The strange collective behavior of cellular slime molds brings us neatly to the next chapter, where the greater good is served by multicellularity.

CHAPTER FIVE
Coming Together— Multicellular Life

Every animal, plant, and fungus we encounter is a community of countless cooperating cells. Different cells within a single organism can take very different forms and have very different jobs, making up distinct tissues and organs. But they all work together for the good of the organism as a whole.

Left *A familiar scene at a familiar scale, dominated by multicellular organisms. A silver-washed fritillary butterfly* (Argynnis paphia) *collecting nectar from flowers of a lavender plant* (Lavandula angustifolia).

Cells and tissues

Every plant and animal is made of several different types of tissue, each one a collection of particular kinds of cells working together. The cells that make up a particular tissue are stuck together with substances that the cells themselves produce and by the specialized junctions that form between them.

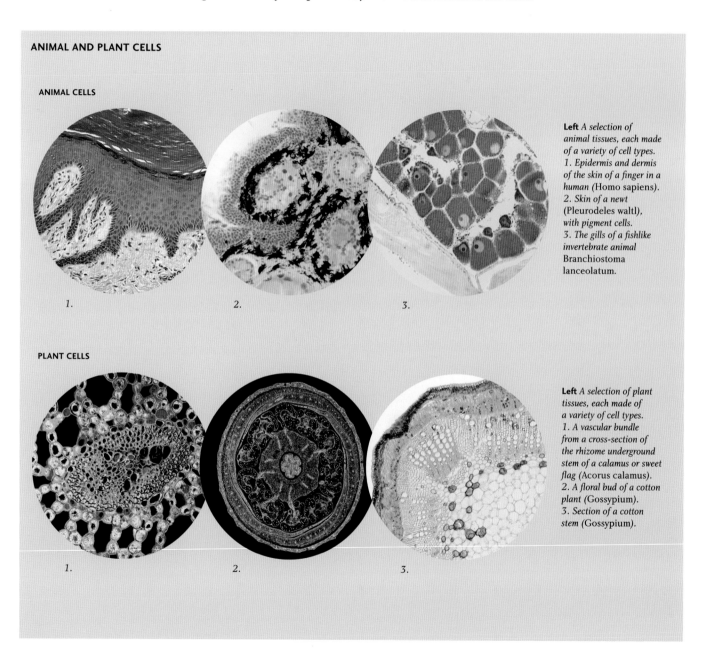

ANIMAL AND PLANT CELLS

ANIMAL CELLS

1.   2.   3.

Left *A selection of animal tissues, each made of a variety of cell types.
1. Epidermis and dermis of the skin of a finger in a human (Homo sapiens).
2. Skin of a newt (Pleurodeles waltl), with pigment cells.
3. The gills of a fishlike invertebrate animal Branchiostoma lanceolatum.*

PLANT CELLS

1.   2.   3.

Left *A selection of plant tissues, each made of a variety of cell types.
1. A vascular bundle from a cross-section of the rhizome underground stem of a calamus or sweet flag (Acorus calamus).
2. A floral bud of a cotton plant (Gossypium).
3. Section of a cotton stem (Gossypium).*

Sticking together

The material between the cells, known as the extracellular matrix, helps hold cells together and give strength and structure to various tissues. In the majority of animals, the two most common substances found between cells are elastin and collagen, both tough proteins that form long fibers. Cells cling onto them, along with other protein fibers that extend from the cell membrane. Collagen and elastin are made from smaller protein units produced inside the cell and released by vesicles bursting open at the cell membrane (see chapter two). Collagen fibers are connected together by strong cross-links—a feature that requires vitamin C. An animal suffering from scurvy, a deficiency of vitamin C, is unable to make viable connective tissue, and as a result joints fail, blood vessels burst, gums ulcerate, and teeth fall out.

While fibrous proteins provide resistance to tension, other compounds called polysaccharides, also present in the extracellular matrix, absorb water to form a gel. This helps to provide support against compression. The amount of collagen and polysaccharide gel present between cells helps to determine the characteristics of a particular tissue type. In some tissues, such as the mucous membranes that line the stomach and the nasal cavity, there is only a little extracellular matrix; these tissues are made almost exclusively of cells. In skin, there is plenty of both fibrous and gelatinous material in the extracellular matrix. In bone, tough collagen fibers are plentiful and are reinforced with mineral deposits; the cells are few and far between. Specialized collagen-producing cells called fibroblasts are relatively plentiful. In cartilage, found in joints or in the outer ear, cells are similarly scarce inside a mass of collagen and polysaccharide gel. In the case of cartilage, the collagen fibers and polysaccharides are produced by cells called chondrocytes.

In plants, the extracellular matrix is mostly made of the cellulose of plant cell walls, with a layer of pectin to help cells stick together. Cells on the outer

Right *Colored scanning electron micrograph of a bundle of collagen fibrils. Collagen is manufactured inside cells as individual molecules and arranges into these long, strong structures after being exported from the cell in vesicles.*

surface of a plant may produce a layer of wax, which helps to prevent the plant from drying out. The extracellular matrix of woody tissue—found in trees and shrubs—contains another compound produced by cells: lignin. In fungi, the extracellular matrix is mostly made of chitin, which also forms the cells' walls. Cells in arthropods (insects, arachnids, and crustaceans) also produce a lot of chitin. It does not only populate the extracellular matrix; a thick layer around the outside of the organism provides a tough exoskeleton. The carapace (shell) of a crab or shrimp and the "fuselage" of a fly are examples of chitin exoskeletons. In some cases, the exoskeleton is in pieces that allow for the animal to grow, but sometimes it is in one rigid piece that must be sloughed off and renewed.

Grabbing hold

In addition to the extracellular matrix, cells in some types of tissue are held together directly, bound by proteins that connect to the cells' cytoskeletons. Such intercellular connections enable cells to form fibers and layered surfaces. The most common type of connection is called a desmosome. These are useful for resisting tearing in tissues that may experience tension. Cells in muscle fibers are joined one after the other by desmosomes, for example, and several desmosomes per cell hold skin cells together in layers.

Some cells need to be held together far more tightly than others. The tubes that make up an animal's digestive tract, for example, have to be tight-knit surfaces that let little or nothing leak out between the cells. The cells of these surfaces are held together by couplings called, appropriately enough, tight junctions. By pulling junctions together at different parts of the cell, surfaces can be made to curve around into tube shapes, as in blood vessels.

Muscle cells in mammal hearts are connected by very sturdy links called intercalated disks. One feature of these links is an opening through which two adjacent cells can communicate or share resources. These openings, called gap junctions, are found in other types of tissue. They are actually transmembrane proteins (see chapter two) that straddle the cell membranes of both connected cells. Gap junctions in heart muscles let the electrical impulses pass quickly between the cells, enabling highly coordinated muscle contractions. Gap junctions between adjacent neurons in certain parts of the brain provide direct electrical connections; most neuron-to-neuron communication happens via signaling compounds called neurotransmitters passing across an actual physical gap between the cells (a synapse, see chapter seven).

Although plant cells are surrounded by a thick, tough cell wall there are also specialized gap junctions between them. These openings—called plasmodesmata—are actually tubes of the phospholipid bilayer formed by the joining of the two cell membranes. A typical plant cell will have a few thousand of these membrane-lined tunnels joining it to its neighbors. Much larger than gap junctions between animal cells, plasmodesmata provide easy passage for all kinds of molecules.

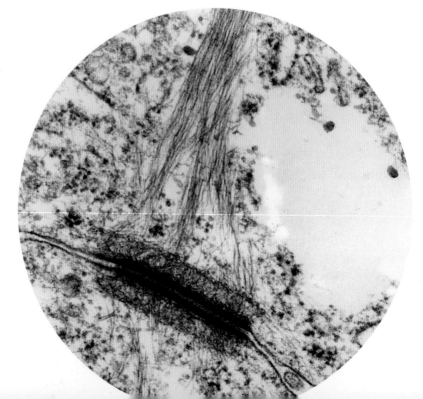

Below *Colored transmission electron micrograph of a section through a desmosome, a junction between cells (dark green). The red-colored fibers extending into the cytoplasm are intermediate filaments called tonofibrils.*

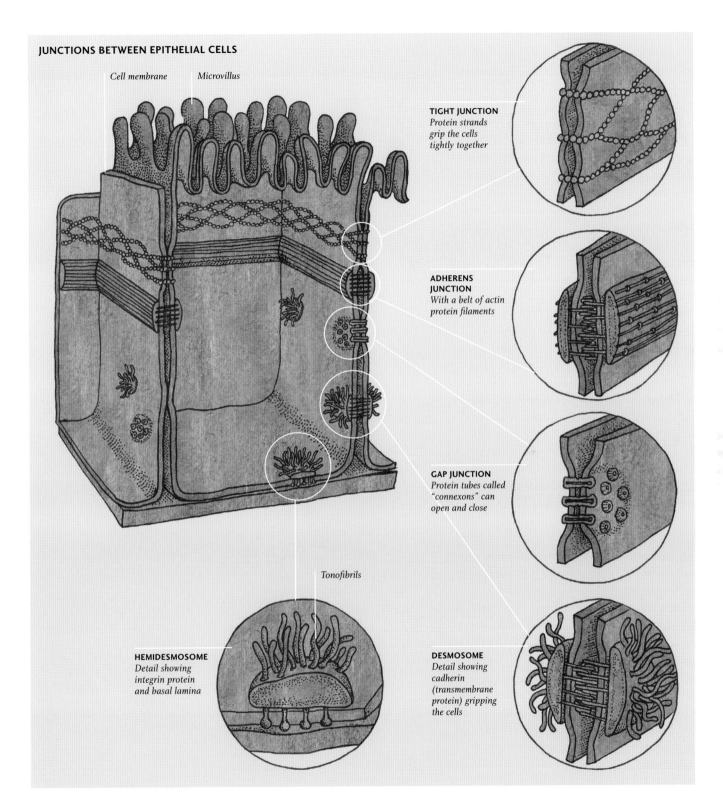

Joining up

Single-celled organisms may be very successful and diverse, but the natural world would be a dull place if they were all that there was. Several times in the history of life on Earth individual cells have clung together and begun benefiting from communal living. Today, the world is full of truly multicellular organisms just as diverse and successful as a world of invisible individual cells.

First contact

The earliest evidence of cells joining together is found in tiny fossils laid down way, way back in time. About 2.4 billion years ago it seems that some cyanobacteria—those photosynthesizing, single-celled bacteria sometimes called blue-green algae (see chapter four)—stayed together instead of separating after they reproduced. The result was an organism made up of many cells connected one after another—like a pearl necklace.

This was almost certainly a colony rather than a true multicellular organism. There are two main differences. First, colonies are formed by single-celled organisms that come together instead of a collection of cells that has grown from a single cell (the cyanobacteria would have passed that test). Second, if a cell is taken away from a truly multicellular organism, that cell dies. Individual cells taken from a colony would be able to take in nutrients, move around, reproduce—and could survive on their own. The cells of the ancient cyanobacteria probably would have been able to do this, so on this basis it would have failed to be classed as a true multicellular organism. Many bacteria form colonies today, albeit mostly in mutually adhering aggregations called biofilms (see box).

Right Colored scanning electron micrograph of dental plaque on a human tooth, showing bacteria and a glycoprotein matrix formed from bacterial secretions and saliva.

Far right A more familiar view of dental plaque.

BACTERIAL BIOFILMS

Colonies of bacteria, called biofilms, share some features with multicellular organisms. There are biofilms clinging to teeth as dental plaque; in rivers, as a slimy coating on rocks; and as "gunk" clogging up sink drains. Inside the body, biofilms can form on surfaces, such as the inside of the urinary tract or blood vessels, and the tight-knit community is more resistant to flushing away—and to antibiotics—than individual bacteria would be. Bacteria within the biofilm reproduce, and eventually the colony releases small clumps or individual cells to form a new film elsewhere.

When bacteria form biofilms, they rely on mutual chemical signals to be "aware" of each other's presence. In particular, they utilize a relatively newly discovered form of communication called quorum sensing. When enough individuals are present nearby—when the cacophony of chemical signals is loud enough—the bacteria alter their metabolisms and produce the material that sticks them all together. This extracellular polymeric substance is a mixture of polymers, nucleic acids, and proteins. As a result of the quorum sensing, bacteria within the new colony form more pili (thin appendages that bind the cells together and enable the exchange of genes), and some of the cells take on different roles from others—some concentrating more on producing the extracellular substance, for example, while others may focus more on reproducing.

1. ATTACHMENT   2. GROWTH   3. DISPERSAL

COMING TOGETHER—MULTICELLULAR LIFE

COLONIAL ORGANISMS GALLERY

Colonies made up of hundreds, thousands, or even millions of individual, single-celled organisms working together are as intriguing as they are beautiful; for how do these cells communicate and orchestrate their communal living?

MERISMOPEDIA
Cell division in the cyanobacterium Merismopedia *is in two axes that are at right angles to each other. The result is that this colonial bacterium forms large rectangular sheets, which are held together by a glycoprotein "glue."*

VOLVOX
Perhaps the most fascinating and most beautiful colonial organism is Volvox, an alga. This spherical organism consists of up to 50,000 cells in a colony-cum-multicellular-organism arrangement.

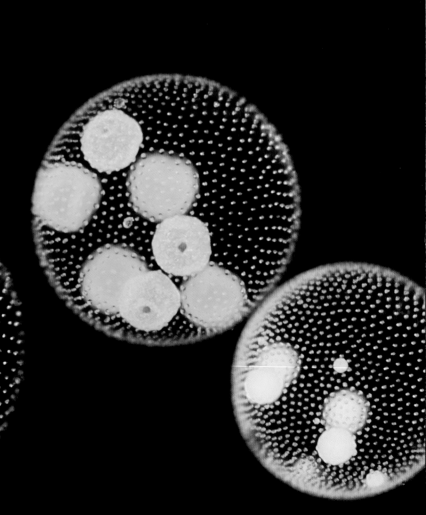

NOSTOC
There are many species in the genus Nostoc, and all exist in colonies. Nostoc pruniforme, shown here, forms grapelike nodules sometimes called "Mare's eggs."

METHANOSARCINA
Some archaeons form colonies, too. Perhaps the most beautiful are colonies of Methanosarcina, which do not look unlike cauliflowers in this colored scanning electron micrograph image.

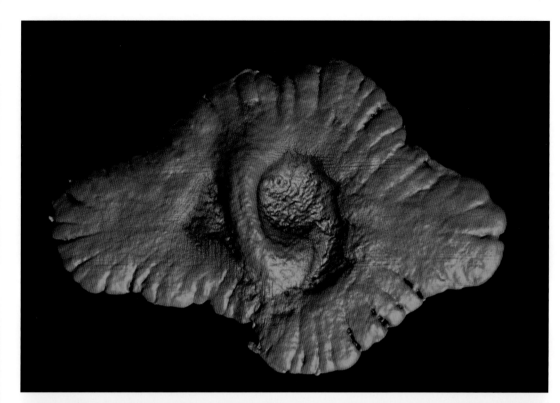

Left *The oldest known candidate for true multicellularity is this thumb-sized mussel-like organism, found in Gabon in 2008 and dated at 2.1 billion years old. These computer-generated images were produced using data from microtomography (a CT scan in miniature).*

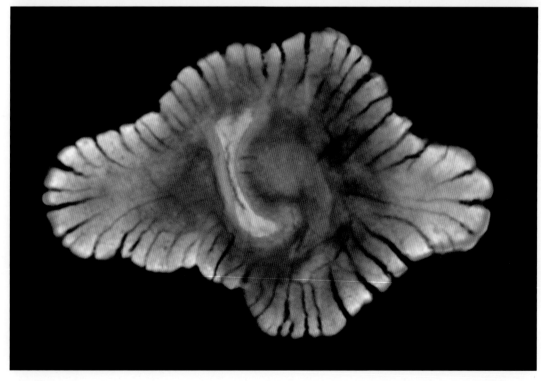

Early life?

Careful studies of the genomes of modern multicellular plants and animals has begun to reveal some details of how the first multicellular organisms evolved, but it was clear before these investigations began, from the study of fossils, that multicellularity has evolved many times in Earth's history. Various examples that seem to have no modern descendants are preserved in the ancient fossil record, including a fairly complex organism that lived around 2.1 billion years ago. Resembling a mussel, this again may have been a large colony, instead of a truly multicellular organism. A billion years later, an early multicellular "creature" named *Grypania spiralis* was drifting around the oceans. This was a stringy, coiled collection of interconnected cells, typically several inches long and was by far the most complex organism of its day.

On Earth today, all true multicellular organisms are fungi, animals, plants, or algae, and most of their lineages can be traced back around 600 million years. (All plants and animals are multicellular, but there are unicellular algae and fungi.) Fungi are soft-bodied, so they rarely leave fossil evidence, but it is probable that the first multicellular fungi were similar to a simple modern fungus called *Chytridiomycota*, which spends some of its life

Below Botrytis, *a fairly simple modern fungus that may be similar to the first multicellular fungi, which lived around 600 million years ago.*

COMING TOGETHER—MULTICELLULAR LIFE

Above *Trilobites were enormously successful multicellular animals. They originated in the Cambrian Explosion, around 522 million years ago, and became extinct some 270 million years later.*

cycle in a single-celled form. The first truly multicellular fungi probably appeared on Earth around 650 million years ago.

Evolution delivered the first animals at about the same time as the first multicellular fungi. But the evolution of animals really took off and flourished a little later during a period of about 20 million years, starting about 542 million years ago. Known as the Cambrian Explosion, this period resulted in the establishment of the lineages of all animals alive today. It is probable that a rise in the concentration of oxygen in the atmosphere was an enabling step. But it is also possible that by this time genomes had become robust and complex enough to produce more sophisticated multicellularity—the genetic toolbox for the general layout of body shape, which is remarkably similar across all animals' genomes today. Before the Cambrian Explosion, all animals were extremely small and simple or lived in colonies. The closest living single-celled relative of animals are simple protists called choanoflagellates. These modern organisms form colonies, and it is probable that the most recent ancestor common to humans and the choanoflagellates, pushing itself around ancient seas with its flagellum, did the same.

Simple plants developed around 500 million years ago—they were organisms whose lineage can be traced back to green algae. The closest living relative

Below *The earliest land plants evolved from multicellular algae that would have looked like this one,* Chara globularis.

Below right *Fossilized cross section of the earliest known "vascular bundles," found in Rhynie Chert, a sedimentary rock deposit in Aberdeenshire, Scotland, dating to 420 million years ago. The evolution of these water-carrying tubes enabled early land plants to grow tall.*

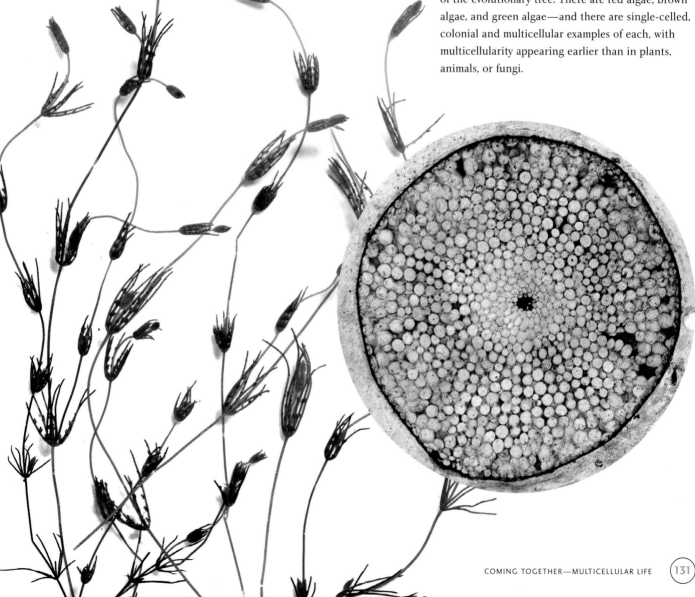

to those first simple plants is Charales, an order of multicellular algae that live in freshwater ponds. In order to gain a foothold (or a roothold) out of the water and to grow higher than an inch or so, plants had to evolve vessels to carry water to all parts of the plant. This vascularization of plants occurred around 420 million years ago. It enabled plants to enter into an arms race, in which the weapon was increasing height, with the result that during the Carboniferous period (360 to 300 million years ago) much of the world's land surface was covered with dense forests of enormous trees. The term "algae" covers a vast set of organisms spread across many separate branches of the evolutionary tree. There are red algae, brown algae, and green algae—and there are single-celled, colonial and multicellular examples of each, with multicellularity appearing earlier than in plants, animals, or fungi.

It takes all kinds

All the cells of a multicellular organism, apart from the sex cells (gametes), carry the same genome. And yet, as an organism grows, perhaps dozens or hundreds of different cell types develop. This differentiation is a key feature of multicellular organisms; it represents the sophisticated division of labor at the cellular level.

On being different

Differentiation is what makes nerve cells, skin cells, and liver cells look and behave so differently from each other. In complex organisms, differentiation is how different tissues are made—a different mix of cell types makes up each tissue. It is amazing that such cellular diversity is possible within a single living thing. Simple multicellular organisms—and even some colonial ones—have basic differentiation.

In the beautiful colonial organism *Volvox*, for example, most cells are somatic—incapable of producing a new organism—but some are germline cells, which will go on to produce the next generation. These germline cells are kept on the inside, safe from harm, and they reproduce to form new colonies inside the parent.

There is a different type of differentiation in the cyanobacterium *Nostoc punctiforme*. This organism exists as a chain of connected cells; most are identical, but every ten or so cells along the chain is a nitrogen-fixing cell called a heterocyst. These cells have a thicker wall, which keeps out oxygen that would otherwise affect the nitrogen fixation. Heterocysts do not themselves photosynthesize—instead they rely on food produced by the other

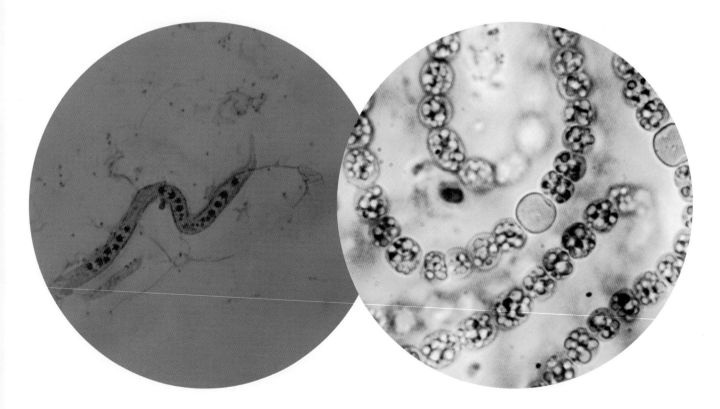

Below left *The simplest animals, such as this Rhombogen found in an octopus's kidney, consist of only a few hundred cells.*

Below Nostoc punctiforme, *a cyanobacterium that is colonial, not multicellular, but that nevertheless exhibits some differentiation.*

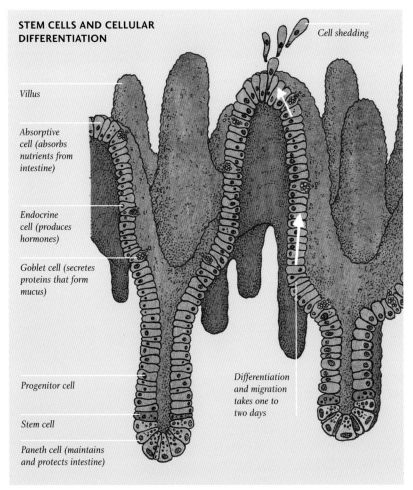

STEM CELLS AND CELLULAR DIFFERENTIATION

Villus

Absorptive cell (absorbs nutrients from intestine)

Endocrine cell (produces hormones)

Goblet cell (secretes proteins that form mucus)

Cell shedding

Differentiation and migration takes one to two days

Progenitor cell

Stem cell

Paneth cell (maintains and protects intestine)

for example, is simply replaced by a younger model after just a few days. Blood cells last a few weeks. Similarly, a cell in the middle of the stalk of a plant has to commit suicide as soon as it is in place (more about the importance of cell death in chapter six), so that its tough cell wall can form part of the system of tubes that brings water from the roots (xylem) or distributes food around the plant (phloem).

It is not much of a life for any of these cells, compared with the self-serving reproductive freedom of the single-celled organism. However, life is driven by the survival of genes not of the organism itself— and certainly not the survival of individual cells. What makes it worthwhile for a differentiated cell to sacrifice its individuality and its life (or the lives of its offspring) is that all the cells of the organism carry the same genome. Their sacrifice is for the greater good. If the organism was a collection of cells with different genomes, there would be much less incentive for a cell to carry out what is literally a dead-end task.

Differentiation is the means by which growing organisms develop and mature organisms repair and replenish their tissues. To see how it works, consider the lining of the intestine, found in all vertebrates, the most rapidly renewed tissue in the body. The internal surface of the intestine self-organizes into tiny fingerlike projections called villi, which greatly increase the surface area through which nutrients are absorbed. The outer layer of each individual villus is populated by four different types of cell, which all have different roles. All these cells are daughters of precursor cells, which are located near the base of the villus. Such precursor cells are themselves daughters of stem cells, the progenitors of all the cells of the outer layer of the villus.

Stem cells are multipotent, which means they have the potential to differentiate into any of the various kinds of cell that a particular tissue needs. Stem cells are key to the development and growth of

photosynthesizing cells, and in a reciprocal arrangement share with their compatriots the nitrogen they have fixed. Despite this somewhat sophisticated differentiation, N. punctiforme is not truly multicellular, because its cells could all exist easily enough on their own; the nitrogen-fixing cells can change back into more general-purpose photosynthesizing cells.

Dying to be part of the team

Unlike a single-celled organism most differentiated cells in a multicellular organism have no chance to reproduce and pass on their genes. Their cell cycle is arrested, so they do not divide but instead die daughterless. A cell in an animal's intestinal lining, multicellular organisms; there is a different kind of stem cell for each type of tissue. They do not carry out vital functions, but they produce the cells that do. Even though stem cells get to reproduce over and over, they, too, are stuck in a dead-end situation: their genome only ever ends up in differentiated cells that are destined to die sooner or later. So no stem cell or somatic cell will ever pass its genome directly to the next generation of the organism as a whole.

Great potential

Differentiation is a one-way street: a fully differentiated cell in the intestinal lining has reached the end of the line. It does not become a precursor cell again. Similarly, that precursor cell can only divide to make cells in the intestinal lining. It does not "de-differentiate" to become a multipotent stem cell. Farther back up the chain of differentiation, however, are cells with much wider potential. These are pluripotent cells, which can differentiate into any type of stem cell for any kind of tissue, so their descendants after several rounds of differentiation could be muscle cells, blood cells, bone cells, or nerve cells.

In animals, pluripotent cells are found only in embryos (see box). But the ultimate progenitors, with even greater potential than these pluripotent cells, have the ability to become a complete new organism. In animals, such cells, called totipotent cells, are fertilized eggs (zygotes) and the cells of a very early embryo. After only a few cell divisions, the totipotent descendants of a fertilized egg differentiate into pluripotent embryonic cells, and the task of building the tissues of a new organism has begun.

In plants, the equivalent of animal stem cells are meristematic cells. They are found in the growing tips of roots and shoots—regions called meristems. In fact, these cells, as well as plant spores and almost any cell in an adult fungus or multicellular alga, are totipotent, like a fertilized egg. They have the potential to produce an entire new multicellular organism. And all it takes to build a new multicellular individual from a totipotent cell is repeated cell division involving mitosis (see chapter two). Well, that and all the amazing machinery of the cell chugging away night and day. And a way of programming cells so they differentiate in the desired way. Oh, and an overall body plan.

Below left *Light micrograph showing cell proliferation inside the meristem tissue in the root of a fava or broad bean plant (Vicia faba). Note the nuclei (stained dark blue) at various stages of mitosis.*

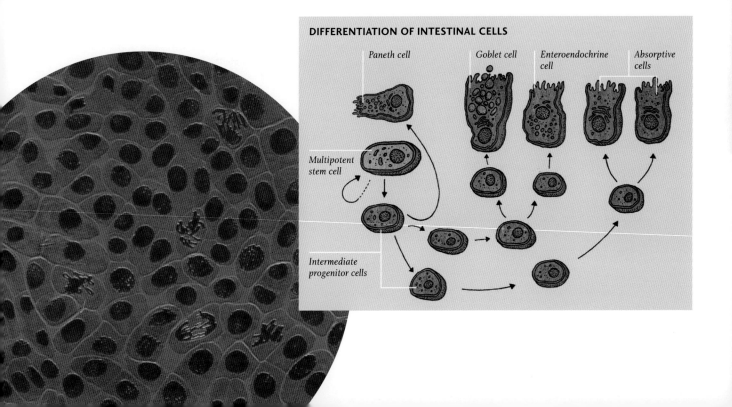

EMBRYONIC STEM CELLS

After a few days of cell division, a mammal embryo is called a blastula. It is made of two distinct cell types: the outer layer is made of cells that will form the placenta, while inside is a clump of pluripotent embryonic stem cells. These cells will divide to form tissue-specific stem cells, which will themselves go on to form the tissues of the new individual. Researchers first obtained and cultured embryonic stem cells from mouse embryos in the 1980s and from human embryos in 1998.

Medical researchers realized that the pluripotency of embryonic stem cells might hold great promise in future medicine. Once removed from inside a blastocyst, they can be cultured, reproducing endlessly to form "seeds" from which new tissue could grow. This tissue could help repair or restore tissues damaged by accidents or disease. It would even be possible to grow entire new organs to order. And simply inserting cultured embryonic stem cells into existing tissues could cause them to differentiate into the right kinds of cell in situ in the body. If the embryo from which the cells are taken carries a patient's own genome then the replacement tissue will be a perfect match for the patient's body, overcoming the problem of the rejection normally associated with tissue donation. To obtain such well-matched embryonic stem cells is a major challenge; those cells only existed in the embryo that has since developed into the very patient who needs the treatment. There are two ways of overcoming this challenge.

The first is to "reprogram" long-differentiated somatic cells back to pluripotency. The resulting cells are called induced pluripotent stem cells, and, once a few are produced, they can be made into a self-propagating cell line. The first clinical trial using induced pluripotent stem cells began in 2014. The second way to produced patient-specific embryonic stem cells is to use somatic cell nuclear transfer—the same approach as was used to produce the embryo that became Dolly the sheep. A cell from the patient's skin or other tissue is fused with an enucleated egg cell, and the resulting cell acts as a fertilized egg with exactly the same genome as the patient. Since this cell has the potential to become a clone of the patient, this technique is filled with ethical problems, and is carefully controlled. The cell divides, and embryonic stem cells are harvested and grown in culture. This procedure was achieved successfully in humans for the first time in 2013.

Above *Colored scanning electron micrograph of a clump of (pluripotent) human embryonic stem cells. These cells have the potential to differentiate into any of the two hundred or so cell types in the human body.*

Right *Phase contrast light micrograph of human embryonic stem cells living in culture.*

COMING TOGETHER—MULTICELLULAR LIFE

Making a body

A new multicellular organism begins as a single cell. Repeated cell divisions and cellular differentiation ultimately result in a complete, functioning multicellular organism—a community of cells with the same genome, ready to compete with other multicellular organisms for food and space and with the mission to pass its genes on to future generations.

Building a multicellular organism

Take one totipotent cell—say, a fertilized egg. Within it lies the potential for building an entire organism; this cell's many descendants will assume all the various roles that the developing multicellular organism requires. For that to happen, the right types of cell must appear in the right position and start doing their allotted jobs. All the instructions needed to make this happen—to unlock the fertilized egg's potential—are inside the nucleus, in the cell's genome. But all the fertilized egg's descendants created by mitosis will also contain that same genome—so what makes a cell more or less potent (less or more differentiated), and what determines its particular characteristics? The answer is genetic expression: which genes within the genome are active and which are not.

Pick out two very different cells in an individual plant—say, pigment-producing cells in the petal of a flower and "guard" cells around the pores in the underside of a leaf. Because they both carry the same genome, the differences between them are

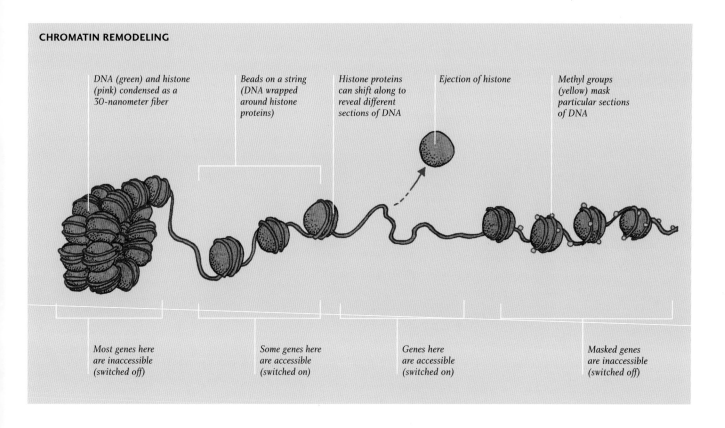

CHROMATIN REMODELING

- *DNA (green) and histone (pink) condensed as a 30-nanometer fiber*
- *Beads on a string (DNA wrapped around histone proteins)*
- *Histone proteins can shift along to reveal different sections of DNA*
- *Ejection of histone*
- *Methyl groups (yellow) mask particular sections of DNA*

- *Most genes here are inaccessible (switched off)*
- *Some genes here are accessible (switched on)*
- *Genes here are accessible (switched on)*
- *Masked genes are inaccessible (switched off)*

COMING TOGETHER—MULTICELLULAR LIFE

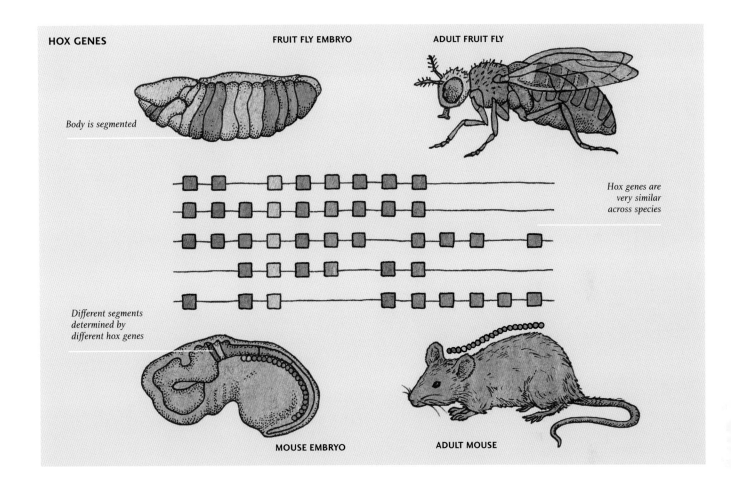

Above All multicellular organisms have sequences of DNA called hox genes (see page 140) that determine what kind of structure goes where. These hox genes are highly "conserved"—they are the same across very different organisms.

down to which genes are being actively expressed. Whether a gene is active or not is determined by physical access to the genes; one set of genes is accessible in the pigment cell, a different set in the guard cell. Imagine an instruction booklet that could be used to build any item of flat-pack furniture in a particular range. Having access to only a certain set of instructions—perhaps covering over the other sets of instructions—would result in someone being able to build either a coffee table or a bookshelf but not both.

If a gene is "accessible," RNA copies can be made as templates for protein building (see chapter two). Most of the time DNA is wrapped around histone proteins, giving it the appearance of beads on a string. The mixture of DNA and histone—the string and the beads—is called chromatin, and the beads are called nucleosomes. Chromatin is further twisted around itself, so a good deal of the actual DNA is naturally hidden away. A number of different enzymes, called transcription factors, directly affect the structure of the nucleosomes (beads) and their position along the DNA (string). This chromatin remodeling is what gives access to particular genes and what decides which genes are turned on and which are off. Genes can also be turned on and off by an enzyme called DNA methyltransferase. This enzyme attaches a cluster of atoms called a methyl group (CH_3) at certain points along the DNA molecule. Where methyl groups attach to the DNA, a gene is switched off.

Right *In the late 19th century the study of the embryology was thrust into the controversy over Charles Darwin's theory of evolution. Ernst Haeckel published drawings showing the embryonic development of several animals, which he believed promoted the idea that all animals are descended from common ancestors.*

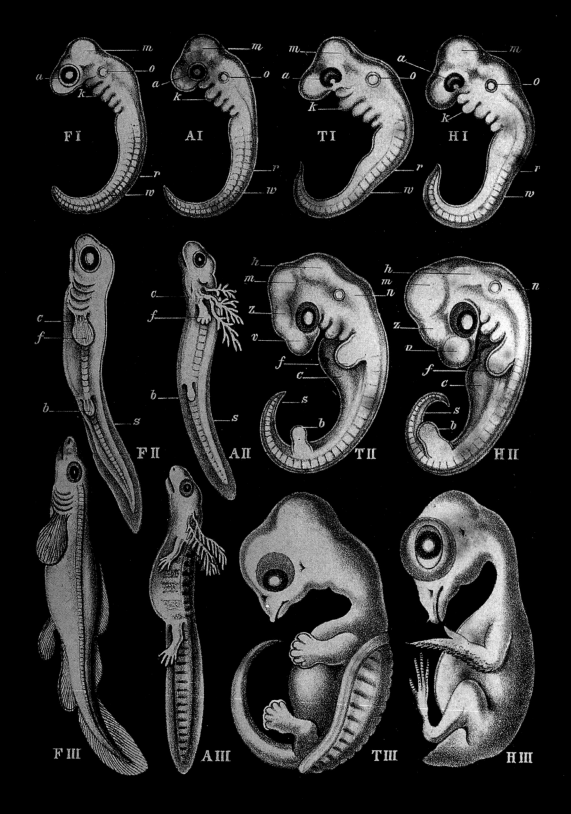

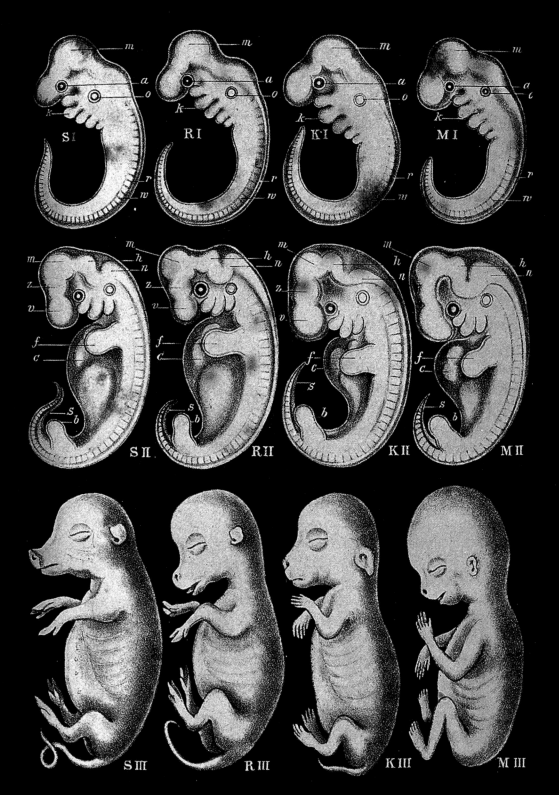

Left *Drawings from Ernst Haeckel's embryology textbook* Anthropogenie *(1874), showing fish, salamander, turtle, chick, pig, cow, rabbit, and human development. Haeckel claimed the drawings were faithful to his observations, but others criticized him for exaggerating the similarities to promote his agenda. Today, there is plenty of solid evidence to support Darwin's theory.*

COMING TOGETHER—MULTICELLULAR LIFE

Short sequences of RNA, called microRNAs, bind to DNA in a similar way to transcription factors and also carry out chromatin remodeling. Molecular geneticists have been able to use microRNAs to reprogram skin cells to become neurons. In the future, this technique could offer treatments for degenerative conditions, just as induced pluripotent stem-cell therapy might—but without the need to "de-differentiate" the skin cells first.

What's the plan?

The genome is a long string of nucleotide sequences, and all it can really do is code for proteins. And yet it manages to carry the overall body plan—the equivalent of architectural drawings for the organism as a whole. The information contained in the genome is organized hierarchically. At the bottom of the hierarchy are genes that code for structural proteins; for signaling proteins, such as hormones, which enable cell-to-cell communication, and for enzymes, which carry out basic tasks. The next level in the hierarchy involves the transcription factors that determine which of those basic genes are active in a particular cell. Above that are further levels of command and control—more transcription factors, which collectively determine the organization of particular tissues and structures. At the highest level is the overall body plan. This determines which of those high-level sets of commands are activated in which part of the organism. These highest-level instruction sets are encoded in sequences of DNA called hox genes. These genes have no direct role in the everyday tasks of the cell. They are not even involved in building body parts. They code for transcription factors that determine what type of body part goes where.

In an animal embryo, a complex mixture of transcription factor proteins, present in different concentrations and produced at different locations, causes the embryo to form distinct segments. Each segment will follow its own development, following a choice of programs selected by the hox genes with cascades of genetic instructions that will build different tissues, organs, and body parts. Make a hox gene active in the wrong place, and the result is a leg growing where an antenna should be or an eye where a leg should be.

All animals have more or less the same set of hox genes. Insert a mouse's hox gene into a fruit fly's genome, for example, and it will work just fine. Just to make it clear that the hox genes are about what body part should be built, but not how, if an eye hox gene from a mouse is activated in a segment of a fruit fly embryo where a leg should be, an eye will grow there but it will be an eye of a fruit fly and not a mouse. Plants and fungi have their own hox-type genes, which determine their body shape in a similar way to animals' hox genes.

One feature of a multicellular organism's development that is important—perhaps surprisingly so—is the deliberate death of cells. So important is this that if it doesn't happen when and where it should, then the life of the entire organism is threatened. The somewhat somber topic of cell death is the subject of the next chapter.

Right *Further evidence of common descent is the fact that all vertebrates have pentadactyl (five-fingered) limbs with the same arrangement of bones. Shown here is a developing black mastiff bat (Molossus rufus).*

Below *A mutant fruit fly (Drosophila melanogaster) with legs where its antennae should be, created by activating hox genes for "leg" in that part of the genome where the "antenna" sequence should be.*

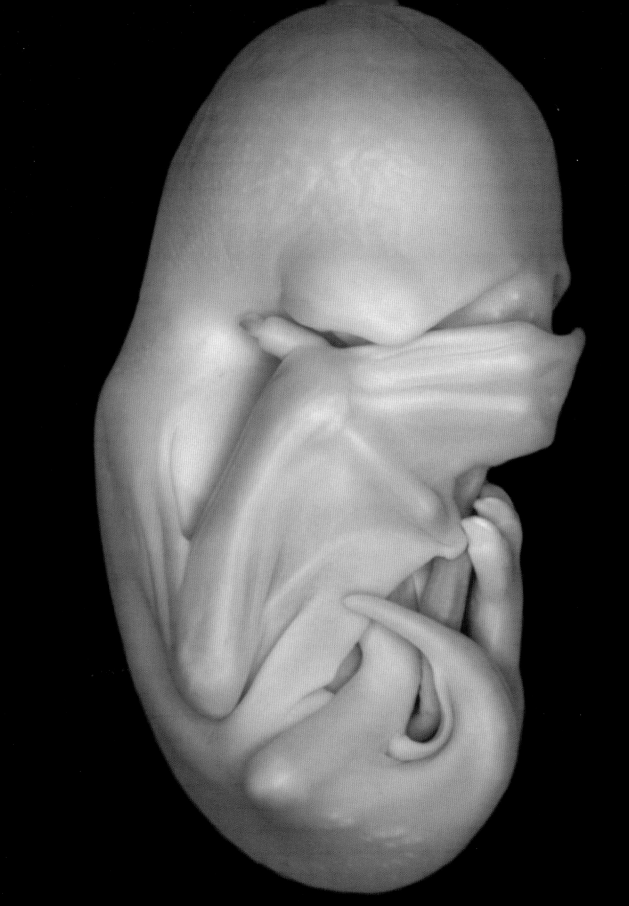

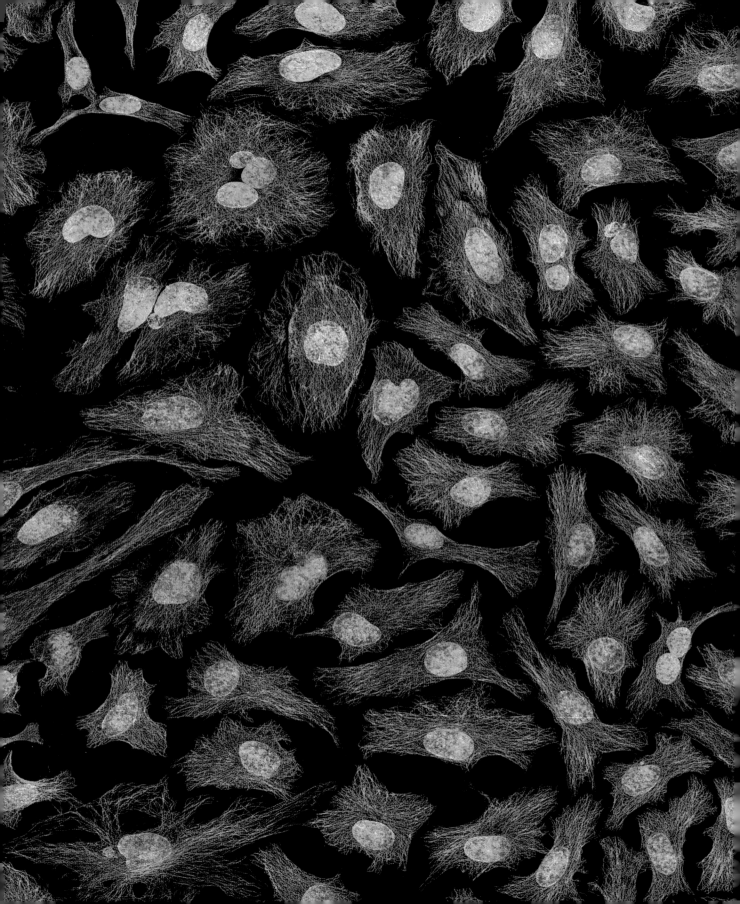

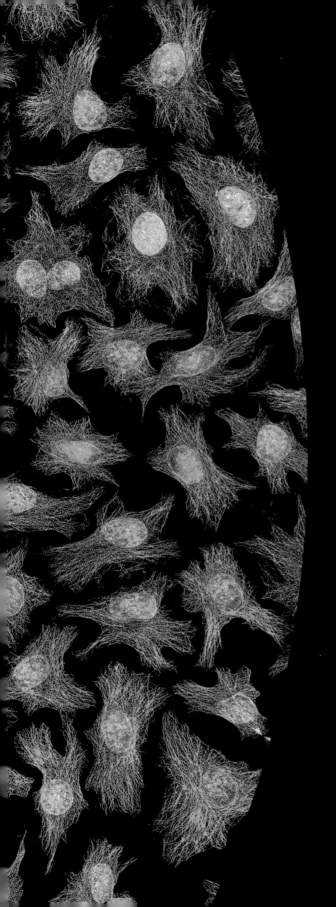

CHAPTER SIX
Life, Death, and Immortality

For cells, life and death go hand-in-hand. In multicellular organisms, for example, huge numbers of cells are deliberately killed by immune systems to keep the whole organism alive and as part of growth and development of the whole. But when errant cells defy death, in an inadvertent bid for immortality, the result may be life-threatening, uncontrolled growth: cancer.

Left *Fluorescent light micrograph of HeLa cells (cytoskeleton is purple, nuclei blue-green). The lineage of these "immortal" cells dates back to a woman called Henrietta Lacks, who died in 1951 after experiencing cervical cancer.*

Order and chaos

Inside every cell, chemistry creates order from chaos—building proteins, nucleic acids, and biopolymers such as cellulose. That's life. Once those processes stop, for whatever reason, a cell's life is at an end. But death is also a beginning and a continuation—for life on Earth is all about cycles and renewal.

Animated debate

Philosophers have long debated the fundamental difference between living and inanimate objects. What makes a bacterium alive, for example, but not a stone? And how can viruses (see box on facing page) be part of the natural world, but not considered as living things? In 1943, the Austrian physicist Erwin Schrödinger gave a series of lectures in which he proposed that one of the key features of living things is the fact that they create order from randomness.

Physicists measure order and randomness using a quantity called entropy. Gases have high entropy, because their atoms and molecules are moving about randomly—far more so than solid objects, whose atoms and molecules can only vibrate around fixed points. Similarly, a large collection of individual atoms randomly arranged has more entropy than those same atoms joined together to form large, structured molecules. Left to their own devices, things tend to become more random—their entropy increases or at least stays the same. An ice cube is a nice, ordered (low-entropy) arrangement of water molecules; put it into warm water, however, and after a while the whole system will be one high-entropy mess of molecules, all moving randomly.

In fact, the inexorable increase in entropy is a fundamental physical law that applies to the universe as a whole. Entropy is increasing overall and eventually, in billions of years' time, all energy will be shared out equally everywhere—a phenomenon referred to as the heat death of the universe. However, although entropy is increasing overall, there are clearly objects or systems in which entropy is decreasing. For example, a freezer can keep an ice cube frozen, but only with an input of energy that will be released at the back of the freezer as heat. The freezer is an "island of negative entropy" but while the contents remain cool, the air around it will heat up and, overall, entropy will increase.

Cells are islands of negative entropy, too. Inside, highly organized processes take place, building up large molecules, such as proteins and nucleic acids, with very ordered structures. In 1964, participants involved in a study sponsored by NASA consulted a group of scientists on how to search for signs of life on Mars. One member of the group was the British ecologist James Lovelock, who suggested looking for "an entropy reduction, since this must be a general characteristic of life." Such is the spontaneous magic of living cells.

Left *Ice cubes have low entropy but, once out of the freezer, they quickly gain entropy and melt. Living things utilize energy from the environment and, through their metabolisms, remain as islands of low entropy.*

LIFE, DEATH, AND IMMORTALITY

LIFELESS VIRUSES

One could be forgiven for thinking that viruses are alive. They are certainly part of life, and they have a lot in common with living things. Each virus particle, or virion, is made of nucleic acid with a protective protein coat; some even have a coating of phospholipid lipids not unlike a cell membrane. The nucleic acid in most virions is RNA, although some contain DNA. In either case, the nucleic acid codes for genes, just as it does in cells' genomes. But, unlike cells, viruses do not have a metabolism nor do they have the capacity to build the proteins for which their genomes code. Instead, they have to rely on the molecular machinery inside living cells.

Although they are not themselves alive, viruses can certainly disrupt the lives of cells. One extremely common class of viruses, called bacteriophages, play havoc with bacterial cells. A phage injects its DNA inside a bacterial cell, and phage virions are reproduced in their hundreds. In order to leave, the new virions have to break out of the bacterial cell. To do this, the phage genome codes for a protein called holin, which makes holes in the bacterial cell membrane, and a protein called lysin, which can break down chemical bonds in the cell wall. Because the phage cannot manufacture these itself, it is the machinery inside the bacterium that unwittingly produces the seeds of its own destruction and lets the newly created virions to escape and infect other cells.

Left *Examples of (nonliving) viruses interacting with (living) cells. 1. Colored scanning electron micrograph of ebola virus virions (red) emerging from a kidney cell (blue) of a monkey. 2. Colored transmission electron micrograph of Middle East respiratory syndrome coronavirus virion on an unidentified cell (note the double-layered cell membrane). 3. Colored transmission electron micrograph of eastern equine Encephalitis virons in a mosquito salivary gland cell.*

LIFE, DEATH, AND IMMORTALITY

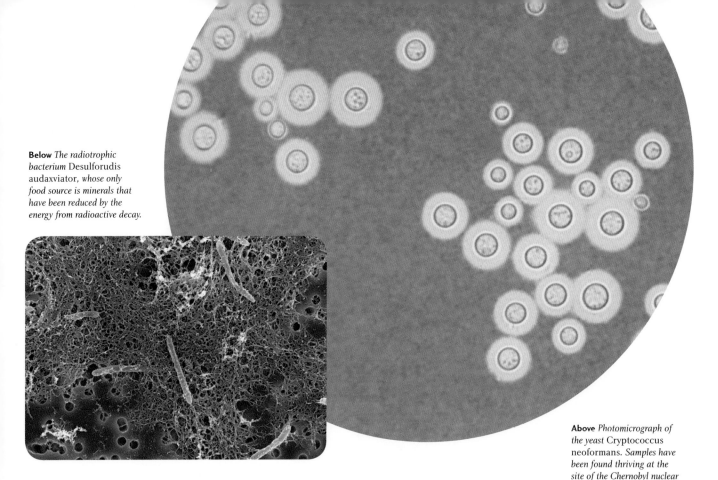

Below *The radiotrophic bacterium* Desulforudis audaxviator, *whose only food source is minerals that have been reduced by the energy from radioactive decay.*

Above *Photomicrograph of the yeast* Cryptococcus neoformans. *Samples have been found thriving at the site of the Chernobyl nuclear power plant, drawing energy from radioactivity.*

Sustaining life

It is theoretically possible for atoms to come together randomly to produce complex compounds. But the probability of, say, a large collection of carbon, hydrogen, oxygen, nitrogen, and phosphorus atoms coming together randomly to form a DNA molecule is vanishingly small. In practice, a cell requires an input of energy to create and maintain order from chaos—just as a freezer does. Organisms whose cells can build complex organic molecules using energy from the environment are known as autotrophs (self-feeders). Most are photoautotrophs, meaning that they use energy from sunlight. Fortunately, the sun will continue to provide free energy for another five billion years or so.

Some autotrophs, however, are chemoautotrophs, which utilize high-energy molecules such as those produced in undersea hydrothermal vents. Deep in a gold mine in South Africa the isolated bacterial species *Desulforudis audaxviator* cannot benefit from sunlight or food from other organisms. It utilizes sulfate (SO_4), not oxygen, as an oxidizing agent in its metabolism, and the energy to produce the sulfate is supplied by natural radioactive decay. One organism that seems able to derive at least some of its energy from radioactivity is the yeast *Cryptococcus neoformans*. A particular strain has been found growing very well around the infamous, highly radioactive ruins of the Chernobyl nuclear power station in Ukraine, destroyed in the world's worst nuclear accident in 1986.

Many living things are heterotrophs. They cannot synthesize their own food and derive the energy they need from ingesting compounds, such as glucose, which have been produced in autotrophs. In other words, they need to eat other cells in order to survive. Generally, then, heterotrophy involves the death of autotrophs—and here is our first example of death being a necessary part of life (true, at least, in a world where so many organisms cannot feed themselves). Fungi are the great recyclers; most of them survive by absorbing nutrients from dead things.

Survival tactics

Most cells will continue to be alive for as long as they are supplied with the required source of energy and the raw materials needed for building complex molecules. If cells are starved of these things, they stop being alive. But they will not die straight away; keep an alga in the dark, for example, and it will use the chemical energy stores it has amassed for as long as it can. The world is an uncertain place, so evolution has written into the genomes of single-celled organisms a remarkable range of behaviors that help them avoid death.

Many are able to move, for example, using flagellated, ciliated, or amoeboid locomotion to follow concentration gradients toward sources of food, light, or oxygen and away from potentially toxic chemical compounds. Nearly all cells also produce heat shock proteins. These fold and unfold other proteins to stabilize the cells against forces that could otherwise cause them to become misshapen and fail to function. The production of heat shock proteins is dramatically "up-regulated" when the temperature becomes uncomfortably high or low, when nutrients are scarce, or, in an aerobic cell, when oxygen is in short supply.

When conditions are just too harsh, some single-celled organisms can enter a kind of suspended animation, slowing their metabolisms to a near-standstill, to reawaken if and when conditions improve. Many species of bacteria can change physically in meager times—into spores. Despite

Below *Molds are multicellular fungi that derive their energy from dead organic matter. The mold shown here is* Penicillium italicum, *which begins as tiny white colonies but develop millions of spores, which are blue-green.*

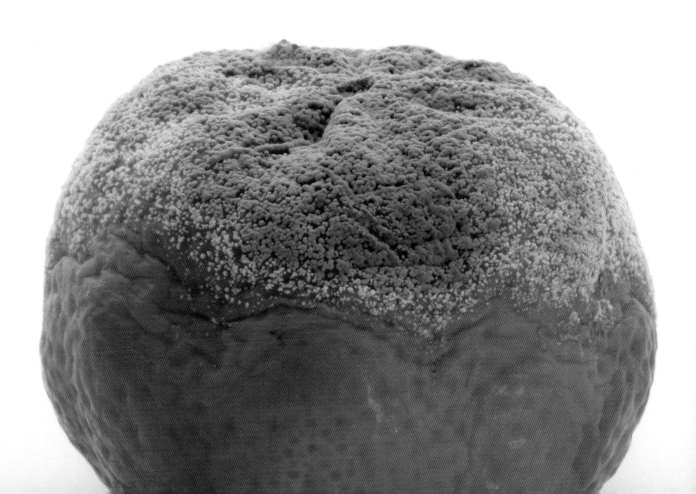

their name, spores are not reproductive structures, they are just minimal versions of the cells on indefinite shutdown inside robust packages.

To change into a spore, a bacterial cell undergoes a remarkable series of changes. First, the single chromosome replicates, and the two copies are divided into two unequal sections within the cell. The smaller section is destined to be the spore. The larger section engulfs it, so that it is now surrounded by two membranes. A wall forms between the two membranes (made of the polymer peptidoglycan, just like the bacterium's main cell wall), and the chromosome of the larger section disintegrates. Next, the spore expels water and builds a sturdy protein coat around itself. Finally, it releases enzymes that break down the outermost original cell membrane. Now the spore is mature and ready to face the world. Bacterial spores can easily remain viable for decades or hundreds of years—and, in certain conditions, for much longer. With careful warming and nourishment, microbiologists have been able to reanimate bacteria that have been frozen in Antarctic ice for hundreds of thousands, even millions, of years.

Dying to be important

Even bacteria grow old as a result of accumulated damage to proteins and other complex molecules. Recently, microbiologists have discovered that when bacteria divide, they partition most of that cellular damage into one of the daughter cells, so that one slightly aged parent cell becomes one older and one

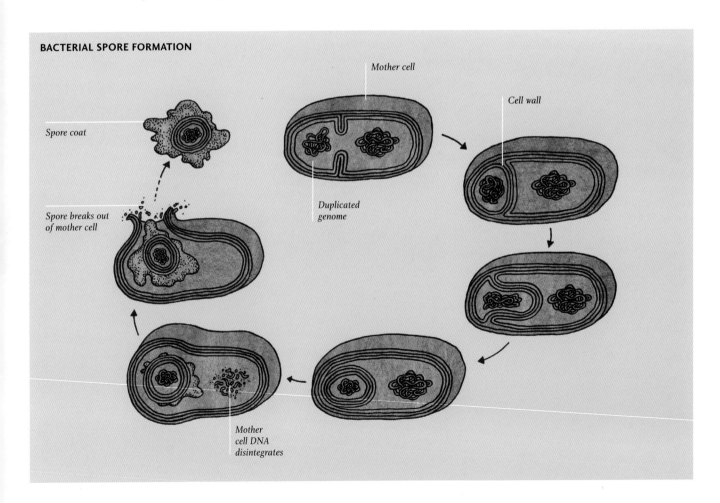

Above *Colored scanning electron micrograph of round spores of the bacterium Myxococcus xanthus. Under normal circumstances the bacteria are bacilli (rod-shape). When the going gets tough they come together and sporulate.*

rejuvenated cell. Eventually, somewhere down the aging line, a daughter cell will be created that carries enough damage that it no longer functions properly and dies without dividing. Meanwhile, the majority of cells created by repeated divisions will be largely free of damage.

Death is important for a number of reasons. It is a central feature of competition for resources, predator–prey relationships, and food chains, all of which sustain life on Earth. As such, death is also an essential part of evolution—and not just the death of individuals but of entire species. An estimated 99 percent of all species that have ever lived are now extinct. If they were not, evolution would not have been able to keep up with the changing circumstances our planet has presented over the millennia—nor would organisms have successfully colonized so many different environments on the planet.

LIFE, DEATH, AND IMMORTALITY

How to kill bacteria

Most kinds of bacteria are either benign or beneficial to us. But there are many that can cause disease—in our bodies or our pets and livestock—or that can cause food to go bad. The death of pathogenic bacteria is important for our own survival—and cunning humans have devised many different ways to despatch these tiny villains.

Kill them before they enter

Fresh food left exposed to the air at room temperature provides an excellent environment for bacteria (and fungi) to live and reproduce. If pathogenic bacteria in food manage to get inside our bodies they can multiply in the intestine. In response, our bodies will try "flushing them out," with very unpleasant consequences. Some bacteria disrupt the tight junctions between cells in the intestinal lining, making themselves a way out into the body at large, where they can multiply and cause more serious problems.

Several food-borne pathogenic bacteria have become household names: listeria, campylobacter, *E. coli*, and salmonella (the species *Listeria monocytogenes*, *Campylobacter jejuni*, *Escherichia coli*, and the genus *Salmonella* respectively). The body's first line of defense against these pathogens is the enzyme lysozyme, which is present in saliva. Lysozyme breaks down bacteria's peptidoglycan cell walls. The second line of defense is the stomach. Gastric juice is mostly concentrated hydrochloric acid with an extraordinarily low pH of 1, which kills most pathogenic bacteria. But many sporulating (spore-forming) species can make it through both these defenses.

So when preparing food for consumption, and especially for storage, it is clearly a good idea to try to kill as many bacteria as possible—or at least restrict their capacity to multiply. The simplest, most time-honored way to do this is to dry food. Water is essential to life, so microorganisms inside dried food either die, slow their metabolisms to the point where reproduction is halted or form hardy but nonreproducing spores. Dried fruits, for example, are extremely concentrated solutions of sugar, because some water is still present. This solution draws water out of any bacterial cells inside the fruit by osmosis.

Salting food works the same way. Refrigerating food slows bacterial metabolisms, so they do not reproduce, but deep-freezing will kill some bacteria, because water expands as it forms ice, and cells may burst open like frozen water pipes. These traditional methods of food preservation have been joined by more recent ones, such as food irradiation—in which food is subjected to X-rays, gamma rays, or electron beams. The energy supplied by the radiation damages DNA, so that bacteria cannot reproduce, and it also produces short-lived, reactive clusters of atoms, called free radicals, that disrupt bacterial metabolism.

Alcohol destroys (bacterial) lives

Water is another thing we absolutely need to put inside our bodies—providing another way for pathogenic bacteria (and waterborne protist parasites) to end up inside us. Modern public-water supplies are remarkably free from pathogens (see box on page 152) although many people are denied such a service, and perfectly preventable water-borne diseases continue to kill millions every year.

Since long before chemical water treatment began people have been purposely fermenting fruit juices. The resulting alcohol beverages are pathogen-free, because alcohol kills most bacteria or keeps them at bay. In solution, alcohol molecules have one end that dissolves well in water and one end that does not,

Right *Three colored electron micrographs of bacteria that cause food poisoning in humans. 1. Escherichia coli strain O157-H7 (transmission electron micrograph, red with flagella), often known as just "E. coli." 2. Listeria monocytogenes (transmission electron micrograph, purple). 3. Salmonella typhimurium (scanning electron micrograph, purple background).*

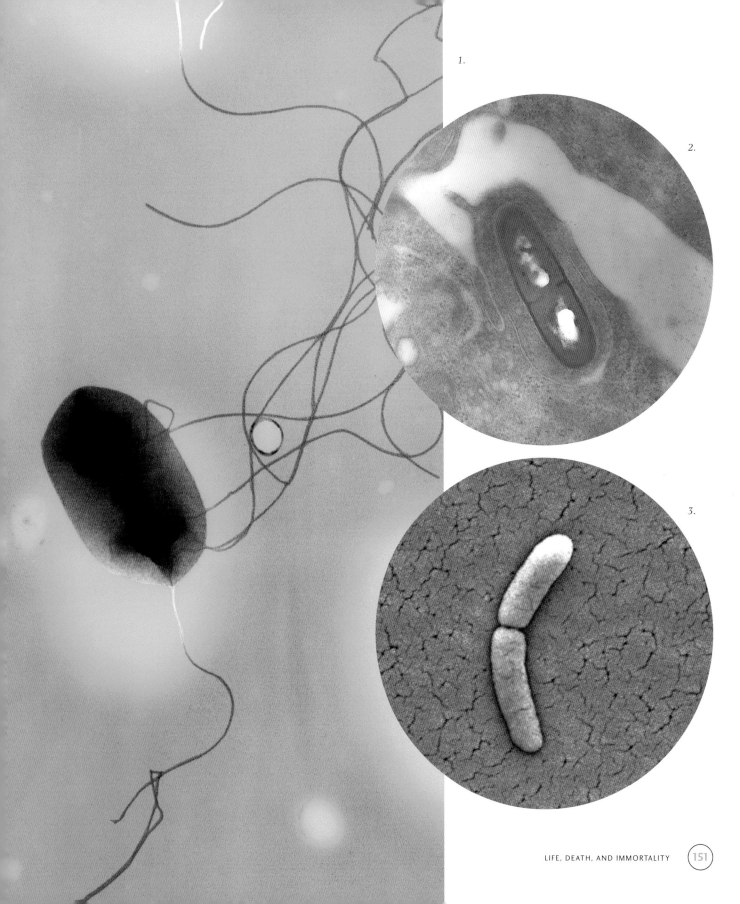

LIFE, DEATH, AND IMMORTALITY

which makes it good at dissolving the phospholipid membranes of bacterial cells. The alcohol present in alcoholic beverages is ethanol (ethyl alcohol), but any type of alcohol has disinfectant properties. Methanol (methyl alcohol) is not quite as effective as ethanol but propanol (propyl alcohol) is most effective of all. Propanol is typically the main ingredient in rubbing alcohol (also known as surgical spirit); it is used to sterilize cuts and grazes and is the active ingredient of hand sanitizers.

Although alcohol is harmful to most microorganisms, some actually thrive in fairly dilute alcoholic solutions. There are bacteria, for example, that derive energy from oxidizing ethanol in wine and beer. The result is ethanoic (acetic) acid, the compound that gives vinegars their sharp taste. Although this acidification is desired by vinegar makers, the process can ruin the flavor of wine and beer. It was in an effort to prevent the spoilage of wine by acetic acid bacteria that the French microbiologist Louis Pasteur came up with a microbe-busting process that now commemorates his discovery: pasteurization.

The importance of sterilization

Today, pasteurization is used in many different parts of the food industry but is most closely associated with milk. Heating to around 160° Fahrenheit (72° Celsius) for about 15 seconds is enough to kill most pathogenic bacteria—and keeping the milk cool thereafter slows the rate at which the bacteria reproduce, giving it a much longer (refrigerated) shelf life. A more extreme process, called ultraheat treatment (UHT), involves heating to about 280° Fahrenheit (140° Celsius) but only for a few seconds. It results in a much longer shelf life still—even at room temperature if the milk is kept in a sealed container. But to kill virtually all bacteria and even bacterial spores—to sterilize a foodstuff—milk and other liquids must be heated to a high temperature for much longer.

WATER TREATMENT

Most modern water supplies contain almost no traces of microorganisms. As well as physical filtering, most contain a chemical disinfectant that kills pathogens. Typically, this is sodium hypochlorite, commonly known as bleach. When hypochlorite dissolves in water, molecules of the compound hypochlorous acid form. These molecules can pass through cell membranes and, once inside, it causes proteins inside to unfold and clump together.

Evolution devised the same strategy a long time ago. Animals' immune systems produce hypochlorous acid as one of their attacks against invading pathogens. In response, bacteria quickly produce a heat shock protein that has evolved specifically to deal with the kind of protein unfolding and clumping hypochlorous acid causes. But, even in a very dilute solution, hypochlorous acid still tends to overwhelm this bacterial defense.

Left *All human life is dependent on water but consuming from contaminated supplies can result in potentially fatal consequences.*

Sterilization is of utmost importance in surgery because when a patient is cut open they are extremely vulnerable to infection. As part of his research into the lives of bacteria Louis Pasteur found that microorganisms could be killed by filtration, heat, and certain chemicals.

Inspired by this, British surgeon Joseph Lister set about revolutionizing surgical practices by spraying phenol (carbolic acid) onto patients and into the air in the operating room. Phenol acts like an alcohol, but it is also acidic, so it is doubly good at killing bacteria. Lister's antiseptic spray saved many lives and in the 1890s an American doctor named Lawson Tait improved surgery farther by insisting that operating rooms and surgeons themselves be made free of bacteria. Tait transformed Lister's antiseptic surgery into aseptic surgery. Today, surgeons scrub up with antiseptic soaps not dissimilar to Lister's phenol, but their surgical instruments are sterilized by heating with steam at over 248° Fahrenheit (120° Celsius) for about 15 minutes in a pressurized chamber called an autoclave. Operating rooms are also fed with air that has passed through "high efficiency particulate air" (or HEPA) filters, which only allow through particles smaller than 0.3 micrometers (0.0003 millimeters)—smaller than the smallest bacteria.

Below *A swab containing povidone-iodine, a solution of polyvinylpyrrolidone and pure (elemental) iodine, a commonly used topical antiseptic.*

Right *Scanning electron micrograph of a HEPA filter. Bacteria and even viruses adhere to the thin fiberglass fibers.*

Defensive strategies

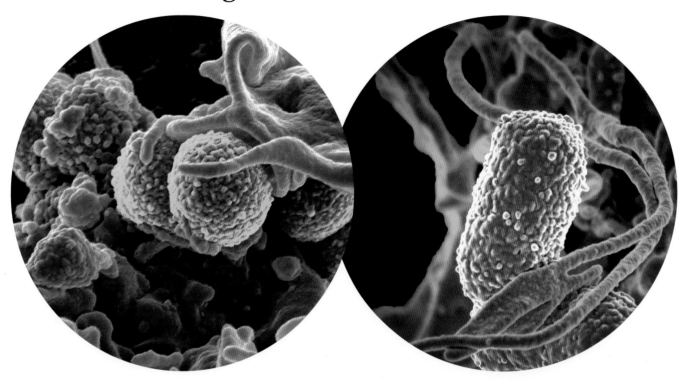

Long before people developed sterile techniques, clean rooms, and antibiotics, evolution had equipped us with an almost impenetrable outer skin—and an incredible set of biological tools for dealing with the pathogenic bacteria and viruses that manage to get inside.

Gaining entry

The outer layer of our skin (the epidermis) is a waterproof and pathogen-proof barrier of dead cells packed with tough keratin protein. But several vulnerable areas of the human body remain: exposed tissues composed of living cells that have no keratin—in particular, the mucous membranes of the respiratory and digestive systems, the conjunctiva of the eyes, and the openings of the genitourinary system. These tissues draw on a range of defensive measures, including the bacteria-busting enzyme lysozyme, which is present in tears and in the mucus

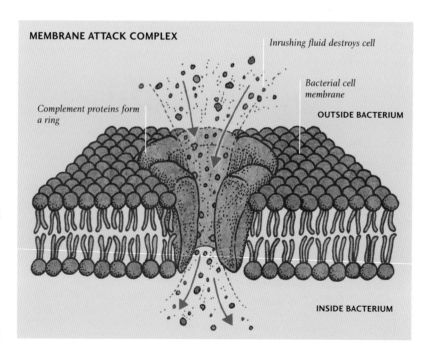

MEMBRANE ATTACK COMPLEX

Complement proteins form a ring

Inrushing fluid destroys cell

Bacterial cell membrane

OUTSIDE BACTERIUM

INSIDE BACTERIUM

Left Colored scanning electron micrographs of (left) Klebsiella pneumoniae and (far left) methicillin-resistant Staphylococcus aureus (MRSA) bacteria being phagocytozed by much larger human white blood cells.

Below Another colored scanning electron micrograph of the phagocytosis of Staphylococcus aureus (MRSA) bacteria, this time colored purple.

produced by mucous membranes as well as in saliva. The cells in the lining of membranes in the respiratory system also constantly wave their cilia back and forth, pushing breathed-in microbes toward the top of the esophagus so they can be swallowed—sending them to their deaths in the acid bath of the stomach.

Despite these excellent border controls, bacterial, viral, fungal, and protist pathogens are able to get through—directly into the bloodstream if the skin or mucous membranes are breached, either by injury or by bloodsucking parasites. Fortunately, most of the time the body's immune system is up to the task of dealing with pathogenic intruders. One of the first reactions is not by immune system cells, but by complement proteins. These compounds are found throughout the body, and they bind to molecules on the membranes of invading bacteria. Some join together to form rings that sit in the bacterial membranes, creating holes. Fluid rushes in through these holes, and the bacterial cell bursts open from within. Some of the 30 or so complement proteins attach to bacterial membranes to label the invaders for destruction. Immune system cells known as phagocytes recognize the labels and begin destroying the bacteria.

Eating bacteria

Phagocytes engulf and digest cells ("phago-" means "eating" and "cytes" means "cells"). When a phagocyte recognizes the complement protein labels, or other compounds present on the bacterial membrane, it encapsulates the bacterium in a vesicle—a bubble made of part of the cell membrane. The new vesicle, called a phagosome, now merges with another vesicle, the lysosome, which contains a potent acidic mixture of toxic compounds. The bacterium is quickly destroyed. When the job is done, the contents of this bag of death (the phagolysosome) are expelled, as it bursts open at the phagocyte's membrane.

There are three main types of phagocyte: dendritic cells, macrophages, and neutrophils. Dendritic cells and macrophages live in the tissues; they are sentinels, waiting for pathogenic bacteria to enter. They are the first responders. While they are gobbling up the offending bacteria, these cells also broadcast chemical signals as part of the inflammation process.

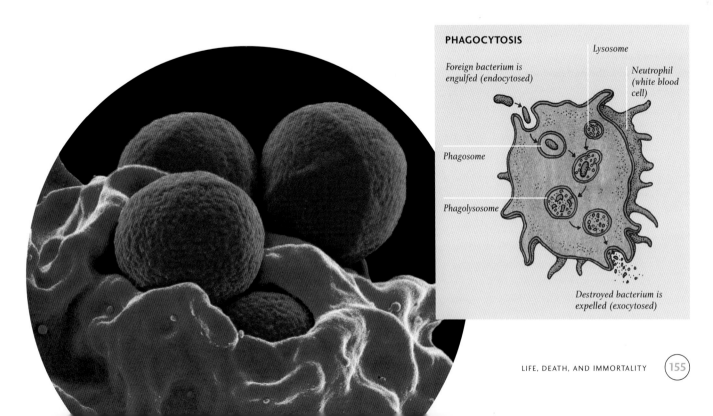

PHAGOCYTOSIS

Foreign bacterium is engulfed (endocytosed)

Lysosome

Neutrophil (white blood cell)

Phagosome

Phagolysosome

Destroyed bacterium is expelled (exocytosed)

LIFE, DEATH, AND IMMORTALITY

One of the effects of inflammation is to loosen the tight junctions between the cells in the lining of blood vessels. The third type of phagocyte, the neutrophils, circulate in large numbers in the bloodstream and, following the inflammatory signals, they squeeze between the loosened cells in the blood vessel walls and out into the tissue. As well as gobbling up offending bacteria, they release cell-killing toxic packages called granules, which burst and help in the bacterial massacre.

Learning to respond

Dendritic cells and macrophages have another function: they help the body learn to recognize specific types of pathogen. They do this by producing "labels" that relate to each pathogen they come across. When these cells destroy a bacterium, they keep hold of fragments of proteins derived from the bacterium they have digested. The fragments of foreign protein become antigens—antibody generators—which attach to a cluster of proteins called major histocompatability complex class II (or MHC class II). The MHC class II complex carries the antigens to the cell membrane.

Now the dendritic cells and macrophages, with their antigen-loaded MHC class II complexes held high (a few dozen per cell), migrate to the lymph nodes. Here they present the antigen labels to T and B lymphocytes, which are involved in the body's learned response (the adaptive immune system). There are various types of T lymphocyte, among them T helper cells, which activate the B lymphocytes, and T memory cells, which exist in small numbers, ready to proliferate if and when the pathogens they remember ever show up again.

Each B lymphocyte produces antibodies specific to a particular antigen. Once taught to produce that antibody (by an antigen-presenting dendritic cell or macrophage), and activated (by a T helper cell), a B cell produces large amounts of its antibody and releases it into the blood.

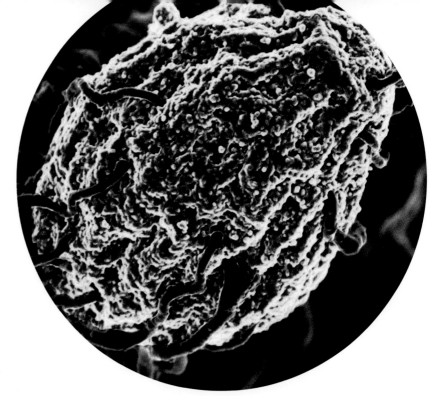

Above *Colored scanning electron micrograph of a macrophage (colored black) engulfing a yeast cell (golden).*

Wearing labels

Antibodies attach to bacteria that bear the corresponding antigen on their membranes, providing a clear target for phagocytes to come and destroy the invaders. An antibody attached in this way (to a bacterial membrane) may also engage complementary proteins. These help to attract phagocytes and can begin the destruction of the bacterium themselves by forming rings that burst the cell open.

While antibodies are attached to bacteria, they may also attach to other antibodies nearby (attached to other bacteria), so that the bacteria are held closely together in custody, awaiting mass execution.

As anyone might guess from the name MHC class II, there is another type of major histocompatability complex, not surprisingly called MHC class I. While MHC class II is only present in macrophages, dendritic cells, and a few other types of immune-system cell, MHC class I is present on nearly all the body's cells. And, like its class II counterpart, this

MHC also lifts protein fragments up to the cell membrane. But these protein fragments are not derived from broken-down bacteria. Instead, they are from proteins that have been manufactured inside the cell itself by the normal DNA protein-building machinery.

In healthy cells, MHC class I complexes present perfectly normal protein fragments, labeling each cell as healthy. When a virus has infected a cell, however, the protein labels are fragments of viral protein, because viruses hijack the protein-making equipment deep inside the cell. At the cell membrane, the MHC class I complexes—in presenting fragments of viral proteins—announce that this cell is infected. So the purpose of MHC class I is to label cells as either "good, leave alone" or "not good, destroy." It is another type of T lymphocyte, called the cytotoxic T cell, which initiates the destruction.

Cell death programme

Cytotoxic T cells constantly check the labels carried by the cells' MHC class I complexes. When they find one that is carrying foreign (viral) proteins, they initiate a remarkable series of events called apoptosis—the equivalent of cellular suicide. The process begins when a receptor on the surface of the T cell meets the MHC class I on the infected cell. The T cell stays close to the target cell and releases granules resembling small lysosomes. Each granule is filled with enzymes, which are introduced into the infected cell. The enzymes begin activating proteins, called caspases, which are already present in the target cell but were previously inactive. These activated caspases activate more caspases, producing a cascade of activity that results in the breakdown of proteins, including the proteins that make up the cytoskeleton. They also activate an enzyme called deoxyribonuclease, which chops the DNA in the nucleus into pieces about 180 nucleobases long.

The dying cell now dissociates neatly into several membrane-bound pieces called blebs, and these are cleaned up by phagocytes. While all this activity is going on inside the dying cell, the cytotoxic T cell that started it all off is busy dividing, making identical copies that will be able to kill other cells infected with the same virus. This is important in stopping a possible viral infection in its tracks and responding quickly if this kind of virus shows up in the future.

Getting rid of unwanted cells

It is important that the caspases and other ingredients of apoptosis are held inside the cell, because there are situations in which cell death can be initiated from within, without the intervention of cytotoxic T cells. In fact, apoptosis claims tens of billions of cells every day in a typical, healthy adult

BLEBBING VERSUS NECROSIS

Intact cell

Blebs

Controlled destruction of cell

Messy death as cell spills its contents

NECROSIS **BLEBBING**

human. For example, it plays a vital part in the development of multicellular organisms and renewal of tissues. In plants, it is apoptosis that clears cells in the xylem and phloem tubes, leaving behind only their cell walls as vessels for the passage of water and nutrients, and it is apoptosis that causes leaves to fall from deciduous trees and spent flowers to die back. Plants also use the self-destructive force of apoptosis as a first line of defense against infection by pathogens, because there is no plant equivalent of complement proteins, dendritic cells, and macrophages. When a pathogen is present, the plant kills off millions of cells in that part of the plant to stop the spread of the disease.

In animals, apoptosis is a crucial part of development. For instance, it is apoptosis that causes tadpoles to lose their tails as they mature into frogs. It can shape tissues and organs, like a sculptor shapes wood or stone by removing material. The hand of a growing human fetus, for example, has continuous flesh around all the fingers; as it develops, tissue is lost between the bones, creating the individual fingers and thumb.

Apoptosis also occurs when irreparable damage has been done to a cell. It is a much tidier affair than unregulated cellular death, which otherwise happens to a damaged cell (see box). Damaged organelles, proteins, and sugars can be replaced easily but damage to DNA is much more serious. Tens of thousands of incidents of damage to DNA happen in every cell of our bodies every day. Each incident is the result of chemical reactions in the warm chaos of the cellular environment or absorption of X-rays or ultraviolet radiation.

DNA damage is distinct from mutation. After a mutation the DNA is still viable—all that has happened is that the sequence of nucleobases has changed. In DNA damage, however, the physical structure of the DNA molecule is compromised. Fortunately, cells are endowed with a range of strategies for detecting and repairing errors. Damage to DNA—or even exposure to the conditions in which damage is more probable—activates a protein called p53. This protein causes other enzymes to increase their DNA repair activity. It puts the cell cycle on pause while repairs are carried out and, if the repair is fruitless, p53 initiates apoptosis, causing the damaged cell to self-destruct.

Right *Colored scanning electron micrograph of a cancer cell undergoing apoptosis (programmed cell death). The cell has broken into several vesicles, known as apoptotic bodies, and is ready to be phagocytozed.*

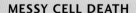

Left *Colored scanning electron micrograph of lymphocytes (yellow) in necrotic tissue. When millions of lymphocytes gather at a site of massive cell necrosis, pus forms.*

MESSY CELL DEATH

Ideally, any cell that dies inside the body does so by apoptosis. During apoptosis, the process of cell death is contained and controlled, and the debris is cleaned up immediately. However, cells can also die as a result of physical trauma or starvation of nutrients or oxygen. In these cases the result is a somewhat more messy and far less regulated end called necrosis. Proteins become misshapen, failing to carry out their duties; wastes accumulate; water fills the cell; the membrane becomes degraded; and, ultimately, the entire cell bursts open, spilling its contents into the extracellular space.

Necrosis typically occurs when a blood clot forms in an artery, preventing oxygenated blood from reaching a particular area. Tissue in the area will become necrotic, as cells there are starved of oxygen. Frostbite is another cause of necrosis, because ice crystals form inside the cells.

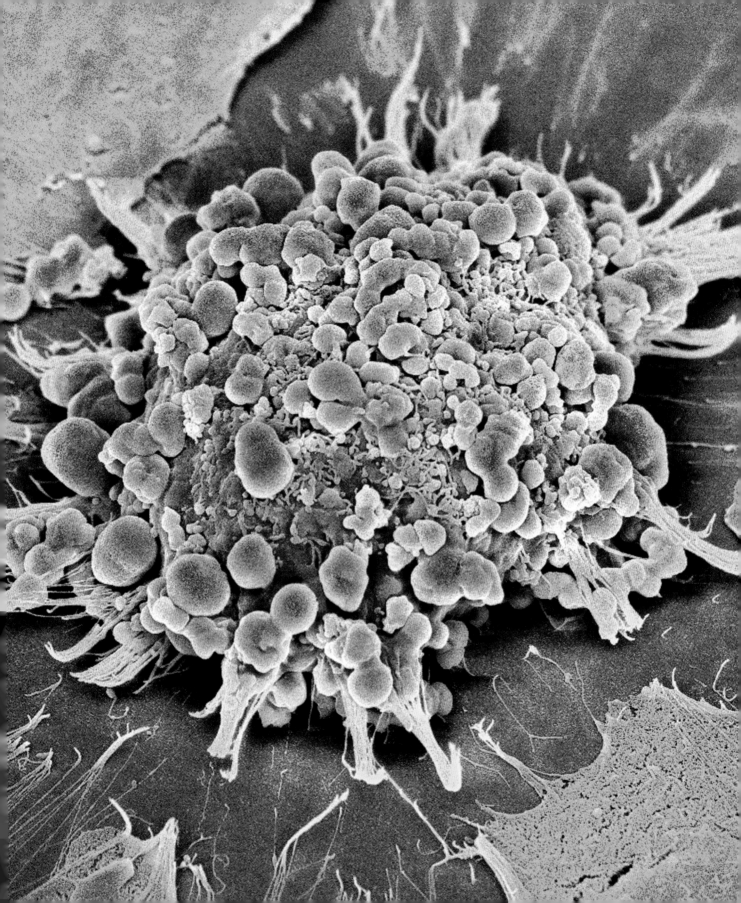

Immortal cells

Certain tissues in a multicellular organism need constant replenishment: new cells to replace old. However, if cells in these tissues proliferate uncontrollably—and without old ones dying off—the result can be life threatening.

Aging cells

In mature animals, there are some tissues in which there is virtually no turnover of cells. In humans, for example, billions of new neurons are formed before birth (and in early childhood in certain areas of the brain) but after that few new ones are added. It's a similar story for the cells of the heart muscle and skeletal muscles. All of these cells have abandoned the cell cycle—they will never divide again—and are, therefore, described as post-mitotic. Because they will never be replaced, even the cell cycles of their precursor cells are permanently paused.

There are problems associated with being post-mitotic. For example, if heart muscle or skeletal muscle becomes damaged, scar tissue forms in its place. In the brain, neurons die every day without being replaced. Just as worrying, perhaps, is the fact that cells age; they accumulate damage in their DNA and their cellular components, some of which cannot be repaired. Instead of undergoing apoptosis and being replaced, damaged neurons and muscle cells have to battle on at less than full capacity. So not only are there fewer neurons in an older brain, but also the ones that remain work less efficiently. (However, with experience comes a plethora of invaluable interconnections that young brains do not possess, so youth is not everything.)

All other cells of the body are mitotic, and the tissues they make are renewed by repeated cell division (at very different rates in different tissues). However, although the label "mitotic" makes it sound like these cells are actively participating in the cell cycle, most will never divide. Instead, these terminally differentiated cells are replaced, by the differentiation of stem cells or precursor cells. They will die without dividing. Dead skin cells and the cells of intestinal linings will be sloughed off and lost forever, for example, while cells in fatty tissue and bone, and glial cells that support neurons in the brain, undergo apoptosis and are phagocytosed and recycled. However, there is a way in which even mitotic cells, in replenishing tissues, can be made to age.

Limiting factors

Differentiated cells that would not normally divide in the body (in vivo) can be made to divide if they are kept alive in culture outside the body (in vitro) and fed with proteins called growth factors. Inside the body, it is the presence of growth factors that causes cells to divide—to encourage them to "un-pause" their cell cycles. If we have a cut in our skin, for example, platelets in the blood release growth factor that encourages stem cells in the skin to begin proliferating, forming new tissue to repair the damage.

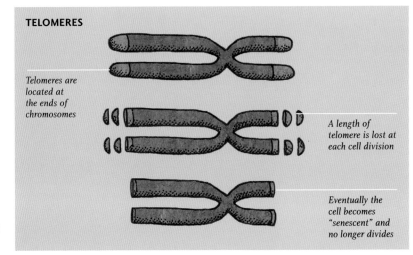

TELOMERES

Telomeres are located at the ends of chromosomes

A length of telomere is lost at each cell division

Eventually the cell becomes "senescent" and no longer divides

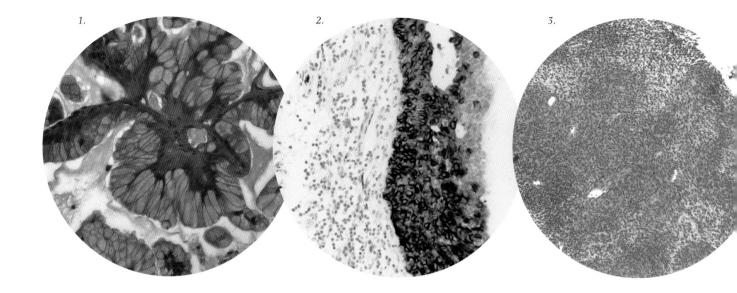

Above Histopathological images of human tissue. 1. Tall (columnar) cancerous cells in the alveoli in lung cancer. 2. Lung cancer, with cancerous cells stained dark brown. 3. Pulmonary sarcoma, a cancer of the connective tissue in the lung.

Cell membranes have hundreds of receptors for growth factors. Before the 1960s, researchers supposed that cells would divide indefinitely if given the appropriate conditions—so that a multigenerational culture of, say, cells from the lining of the lung, could be kept alive indefinitely. But in 1961, American biologist Leonard Hayflick discovered that there is a limit to the number of generations cells survive. Stem cells—which, in culture, divide but do not differentiate—can produce about 50 generations before the cells stop dividing. They have then become old, or senescent.

In the 1970s researchers discovered the reason for cellular senescence—for the 50-or-so-generation "Hayflick limit." They discovered features called telomeres on the ends of each chromosome in eukaryotic cells. Telomeres are often compared with the hard plastic pieces wrapped around the ends of shoelaces that protect the laces from fraying— although we also have to imagine cutting back the plastic each time we tie our shoelaces. Every time a chromosome is reproduced, a small length of DNA at the ends is left out of the replication process. The telomeres are themselves made of DNA—but it is DNA that is noncoding and can,

therefore, be sacrificed with no adverse effects. So the telomeres act as a kind of buffer that protects the all-important coding DNA. They grow shorter with each division, and eventually, after many generations, they disappear; the cell becomes senescent and either dies or no longer divides. Telomeres are of different lengths in different species; in human cells, they are long enough that a chromosome can survive up to about 50 divisions.

One of the benefits of telomeres seems to be that they help to avoid an uncontrolled proliferation of tissues—for there is a fine balance to be struck in replenishing tissues made of mitotic cells. When cell division in a tissue continues indefinitely for some reason the daughter cells will develop into unwanted and dangerous lumps, or tumors. Each would-be tumor starts with a single errant cell; the Hayflick limit and the shortening telomeres can limit the number of generations that cell can produce, limiting the proliferation of cells and stopping the cancer in its tracks. However, the cells of cancers that become established have a feature that restores telomeres to their original length with each division—something that makes them able to proliferate unchecked.

Some types of cell are unaffected by the shortening of telomeres, thanks to an enzyme they produce called telomerase. This enzyme repairs the cells' telomeres back to their original length after each cell division. Normally, the only cells that produce telomerase are embryonic stem cells, from which all the tissue types can be generated, and germ cells, such as eggs and sperms, which have the potential to form a new individual. It is understandable for these cell types to reproduce unhindered. Because they do not grow old—because their reproductive lives are not limited—they are sometimes referred to as immortal cells. More than 90 percent of cancer cells express telomerase, too. So cancer is typically a disease of immortalized cells in which the telomere shortening safeguard has been circumvented. Immortalized cell lines—either embryonic stem cells or cancer cells—can be kept in cultures indefinitely and are incredibly important in modern cell biology.

Cells have other safeguards besides telomeres to prevent the formation of tumors. For example, thanks to contact inhibition, they will not divide if they are touched on all sides by other cells. Even in vitro, contact inhibition stops cells proliferating once the culture dish is totally covered by a layer just one cell thick. Then there are proteins such as p53, which kick-start apoptosis when DNA damage is irreparable. These proteins are called tumor suppressors, because they rid the body of errant cells that might otherwise proliferate wildly. Unfortunately, however, there are occasions when all of these safeguards fail. Cancer is generally the result of damage to the DNA that codes for tumor suppressors, contact inhibition, and the inhibition of telomerase expression. This damage is caused by certain chemical compounds that produce reactive compounds inside the body, by exposure to external radiation, and also through copying errors during mitosis (see chapter three).

Cancer crops up in proliferative tissues—those tissues that contain mitotic cells. Breast cancer, for example, normally begins in the lining of the milk ducts, a layered tissue that is constantly replenished. Other common cancers are found throughout the digestive system, the respiratory system, skin, bone, and the genitourinary tract; in the brain, tumors commonly involve (mitotic) glial cells. Blood cells proliferate constantly, too, in bone marrow and in lymph glands, and there are a number of cancers involving these cells. Some kinds of cancer cell produce their own growth factors and also have many more growth-factor receptors in their membranes.

Fighting back

Many tumors are benign: their growth ultimately slows down, and they do not interfere with any vital processes. Others are malignant: they continue to grow, even forming their own dedicated blood supply. They may eventually release pieces into the bloodstream that form into new tumors elsewhere, a phenomenon called metastasis. The first line of attack against a malignant cancerous tumor is normally surgery. If the tumor is discovered

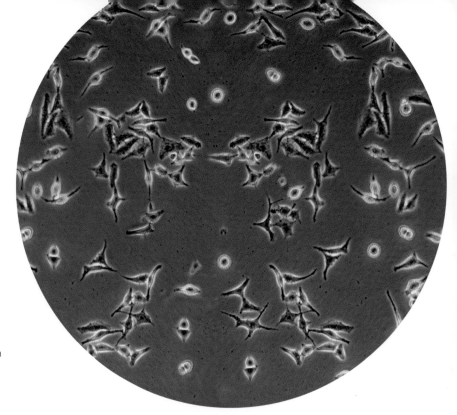

Above *Phase contrast light micrograph of prostate cancer cells in culture.*

early enough, before it has invaded too much tissue or metastasized beyond its original site, then removal will put an end to the uncontrolled proliferation—that will be that. Unfortunately, tumors are not normally discovered until surgery is unlikely to be completely effective. Beyond surgery, there are two standard approaches to getting rid of cancer: radiation therapy and chemotherapy.

Radiation therapy involves the use of ionizing radiation—beams of rays or particles that knock electrons from atoms, disrupting chemical bonds. Sometimes high-energy protons or electrons are used. Because they do not penetrate very far before being absorbed, normally they can only be used for tumors that occur near the surface, such as those in skin cancer. More commonly, the radiation used is gamma rays, emitted by radioactive substances, or high-energy X-rays from X-ray machines. Gamma rays are more penetrating and so can be used to treat cancers discovered deeper in the body. Because normal cells have inbuilt DNA repair mechanisms, they are usually able to recover, or at least undergo apoptosis and be replaced by differentiated stem cells.

The general approach of chemotherapy is to target features found only in cells that are dividing—that way, most of the patient's normal cells are unaffected. Chemotherapeutic medicines interfere with the cell cycles of proliferating cells or induce apoptosis in them, typically by inducing damage in DNA or RNA. However, not all medicines take this approach. One of the most successful breast-cancer medicines, paclitaxel, works by inhibiting the breakdown of microtubules, an essential feature of dividing cells. A medicine called gefitinib, used in treatment of certain types of lung and breast cancer, inhibits cells' growth-factor receptors, slowing down a tumor's rate of growth. Interestingly, when this medicine was first trialed, in the early 2000s, it was only around 10 percent effective. It turned out that only about one in ten lung cancers involves cells with overactive growth-factor receptors. When the medicine is given to those patients, its success rate is much higher. In patients whose cancerous cells proliferate for other reasons, gefitinib had no clinical benefit. By studying the genomes of the cancer cells, doctors can target the treatment, increasing the likelihood that cell proliferation—and therefore dangerous metastasis—can be halted. This promising approach is called pharmacogenomics.

Cancer is difficult to stop in its tracks because of the incredible complexity of the molecular-scale processes that go on inside cells—the very processes that create order from chaos and keep us alive. That complexity is what makes us who we are, what leads to the incredible diversity in our cell types that keeps our amazing bodies working.

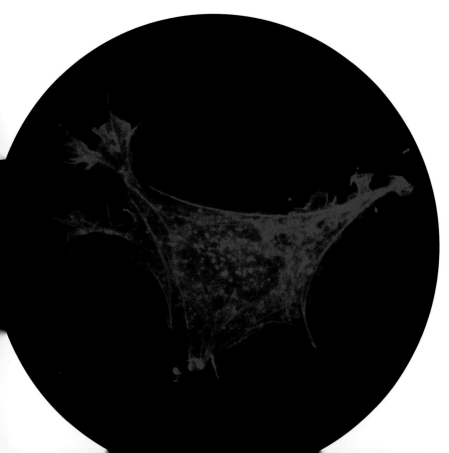

Below *Fluorescence light micrograph of metastatic breast cancer cell invading the extracellular matrix. The red color shows up part actin microfilaments in the cytoskeleton.*

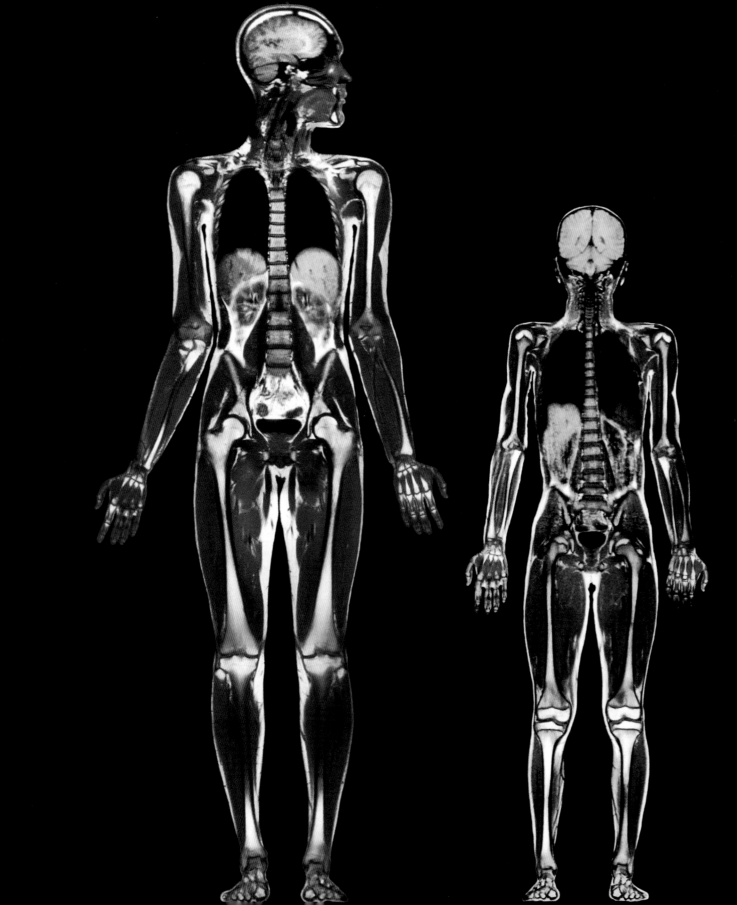

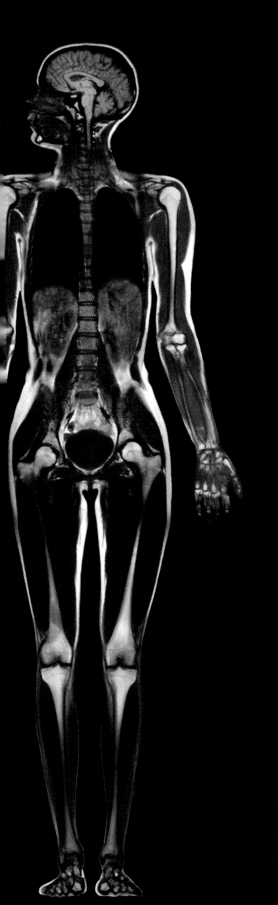

CHAPTER SEVEN

Taking in the Cytes

Humans are complex multicellular organisms, so there is remarkable diversity and specialization among our cells. Our bodies manufacture around 200 different cell types, including adipocytes, erythrocytes, and neurocytes. This last chapter, drawing on what we know about cell biology, takes a tour of the scientific marvel that is the human body—while enjoying the cytes.

Left *Colored magnetic resonance imaging (MRI) scans of a man (left), woman (right), and a nine-year-old boy (center). It is possible to pick out various organs and tissues, such as the lungs, bones, and brains.*

Cells of the human body

Most people learned something about cells at school but perhaps never realized exactly how important, interesting, and stunningly beautiful they can be. All the cells featured in this chapter fall into these categories. Knowing what goes on inside them only makes them more fascinating and even more beautiful.

Erythrocytes (red blood cells)

We begin with a classic and instantly recognizable cell: the disk-shape, biconcave (indented on both sides) red blood cell. In one sense, the red blood cell is not actually a cell at all. It has no nucleus and, therefore, no chromosomes—and is technically called a cell fragment. All blood cells, red and white, begin as multipotent blood stem cells called hemocytoblasts. Several divisions and differentiations result in the formation of immature red blood cells called normoblasts, and it is at this stage that the nucleus is expelled. Two more divisions later, and the tiny, enucleated red blood cell is circulating and maturing, ready to do its vital job of carrying oxygen around the body.

Most of its mass (95 percent of the dry weight) is hemoglobin protein, each molecule of which has four iron atoms at its center. Each iron atom can hold one oxygen molecule. It is to these iron atoms that the oxygen binds after it has diffused across cells in the lining of the lungs. Each hemoglobin molecule can hold four molecules of oxygen, and a typical red blood cell contains about one-quarter of a million molecules of hemoglobin, so each red blood cell can hold about a million oxygen molecules.

A teaspoon of human blood contains about 25 billion red blood cells. Over two million are created every second in a healthy adult human body. And every second the same number are destroyed, ready for recycling in the spleen, each one having lived for about four months. One of the breakdown products of all that hemoglobin is a yellow compound called bilirubin. It is responsible for the yellow color of nearly healed bruises and the brown color of feces as well as the background yellow color of urine.

Red blood cells account for about one in four of all the human body's cells, but only about one-tenth of the body's mass. This is because they are also among the smallest of our cells—about 200 of them lined up next to each other would cover about one twenty-fifth of an inch (1 millimeter).

Right Colored scanning electron micrograph of red blood cells in a vein of a human liver.

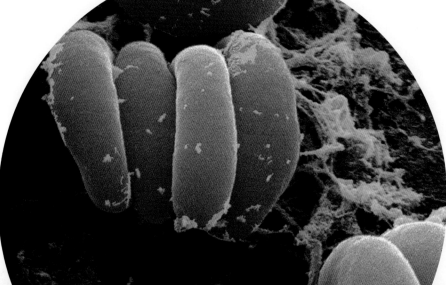

Right The biconcave shape of red blood cells is evident from this colored scanning electron micrograph.

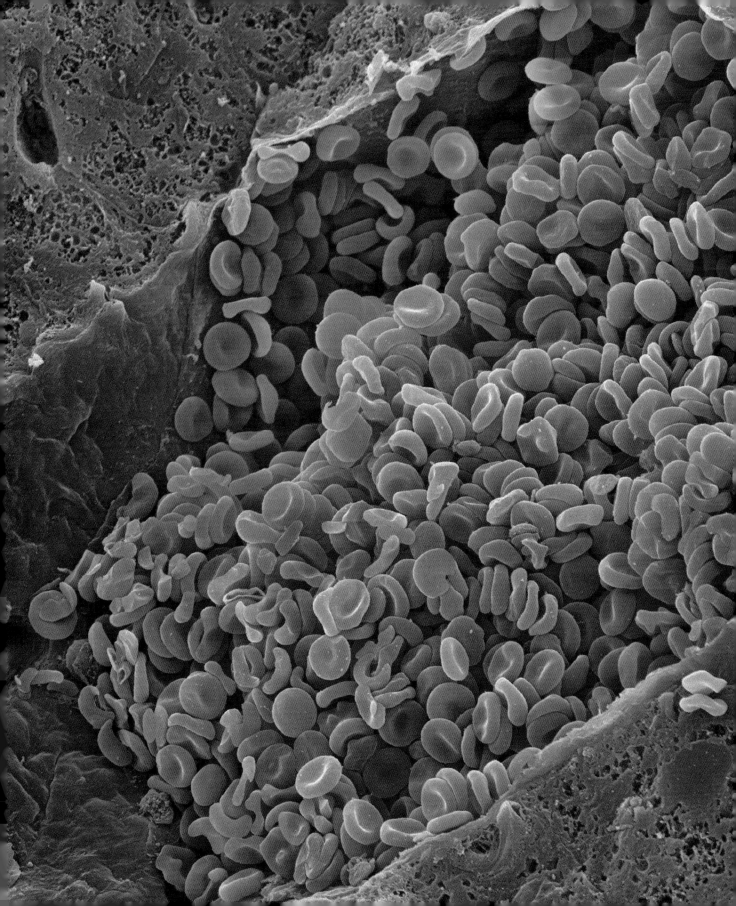

Right *Light micrograph, its field of view dominated by red blood cells with several platelets (smaller, dark stained).*

Left *Colored scanning electron micrograph of an activated platelet (thrombocyte). Activated platelets secrete chemicals that cause the formation of a fibrin mesh, which traps platelets and red and white blood cells, forming a clot.*

Thrombocytes (platelets)

Another enucleated cell fragment found in the blood is the platelet, or thrombocyte. Platelets have a key role in the formation of clots. Without them, even a small injury would not stop bleeding and could become life threatening. They are attracted to any tear or damage in a blood vessel. When the platelets arrive, they are in their inactive, lens-shape form. As they become activated, they quickly change shape, as growing microtubules in their cytoskeleton push out waggly arms. The platelets adhere to the extracellular matrix around the damaged part of the blood vessel and to each other, forming the beginning of a clot, or thrombus.

The activated platelets now release the contents of granules they have stored within their cytoplasm. The contents include calcium, which is an essential ingredient in a cascade of reactions between soluble proteins in the blood, with the result that strands of the insoluble protein fibrin form. The fibrin strands hold the platelets, red blood cells, and the edges of the tear together as a clot. Some of the platelets' granules contain growth factors that encourage the cell division required to heal the broken blood vessel. Activated platelets even release a bactericidal compound that will destroy certain types of bacteria if they enter the wound site.

Melanocytes (pigment cells)

Melanocytes produce the pigment melanin. They are found in hair follicles, and melanin is responsible for hair color. These cells are active throughout life, adding color to hair as it grows. Their activity slows with age, which explains why hair normally grows gray in later life. Melanin evolved long before humans did and is found in many other animal species. For example, it is one of the main pigments in bird feathers, many fish scales contain it, and it also forms part of the immune systems of insects.

However, apart from determining the color of our hair, melanocytes are most important in our skin. They are found at the bottom of the epidermis, the top layer of the skin, and melanin is the body's natural sunscreen; it is very good at absorbing ultraviolet-B radiation (UVB), a particular part of the ultraviolet spectrum that causes damage to DNA.

The human species evolved in Africa, near the equator, where UVB levels are high, and all early humans had dark (melanin-rich) skins to protect them. Some of the early people who migrated out of Africa tens of thousands of years ago ended up far from the equator, where UVB levels are much lower—so low, in fact, that dark skin put them at a disadvantage. A certain amount of UVB is needed for the body to manufacture vitamin D, which is essential for the uptake of calcium. So in higher latitudes, natural selection favored those individuals whose melanocytes made a little less melanin, giving rise to the wonderful range of skin tones present

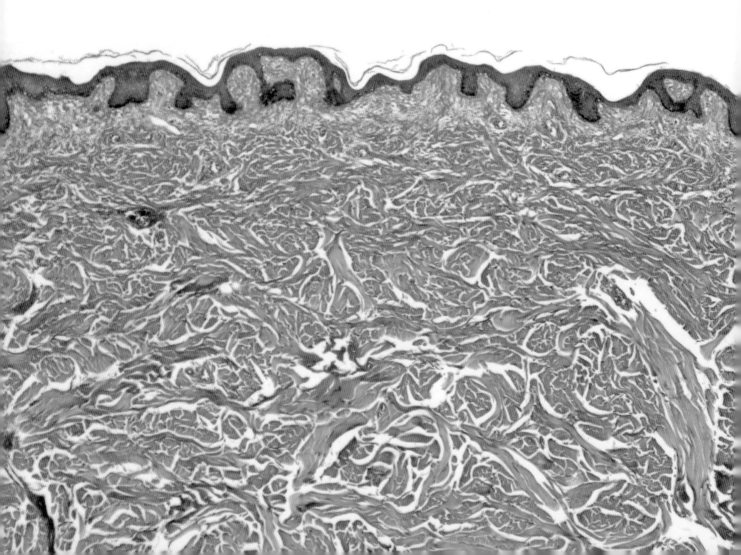

Below *Light micrograph of a section through the dermis and epidermis of a person with fairly dark skin. The pigment-producing melanocytes are visible in a brown layer at the bottom of the epidermis.*

throughout the world today. Surprisingly, perhaps, there is very little difference in the number of melanocytes between any two people; different skin tones are caused by different amounts of melanin per melanocyte.

Inside melanocytes, melanin production takes place in membrane-bound organelles called melanosomes. These organelles are mobile; they can be transported to the ends of the cell's extended branches, which are known as dendrites. In the epidermis of the skin, melanocytes even donate melanosomes to neighboring keratinocytes, the most common epidermal cell. (The keratinocyte gathers the melanosomes by engulfing the ends of the melanosome-packed dendrites.)

Below *Macro photograph of a melanoma—uncontrolled, cancerous proliferation of melanocytes on a human arm.*

Above *Light micrograph of an artificially grown culture of melanocytes, showing the cells' dendrites (branches), along which pigment-containing vesicles can be transported.*

TAKING IN THE CYTES

Adipocytes (fat cells)

Fats are essential for building cell membranes, for high-density energy storage, and, under the skin (subcutaneous), for providing insulation against cold and protection against knocks. Fat is stored in fat cells, also called adipocytes or lipocytes. When we eat more calories than we burn the excess ends up here, as fat, even if the excess is sugar, because the liver easily converts sugars to fats. Inside an adipocyte, the fat is stored as one or more large, semiliquid globules.

These globules dominate the cell's volume, and its nucleus is pushed against the cell membrane. If we gain or lose weight as fat, the number of adipocytes remains the same—each cell simply grows or shrinks to take on or lose fat. In situations of extreme weight gain, however, extra adipocytes are created. Fat cells are mitotic: they are routinely replaced by differentiating precursor cells called lipoblasts. Each fat cell lasts about ten years—so about one in ten is replaced each year. When they are replaced, they are destroyed and dismantled via apoptosis, and the fat globules transferred to the replacement cells.

Fatty tissue is simply made of fat cells bound together by connective tissue, mostly collagen. There are two types of fatty tissue: white fat and brown fat. The cells of brown fat are different from those in white fat. Brown fat cells contain a lot of mitochondria (they are what give the brown color) that burn the fat to produce heat. Newborns and young children have a lot of brown fat around their trunk to protect them against hypothermia. Adults have far less.

Fat cells produce a hormone called leptin, which has a powerful effect on appetite; the more leptin circulating in the blood, the more satisfied a person feels. The innate desire for high-energy food, left over from a time when food was hard to come by, is a stronger influence on appetite; the easy availability of sugary and fatty foods means that more and more of the world's adipocytes are cramming in fat.

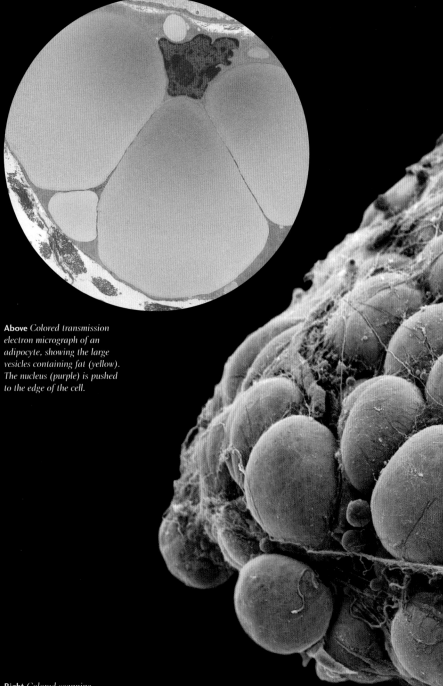

Above *Colored transmission electron micrograph of an adipocyte, showing the large vesicles containing fat (yellow). The nucleus (purple) is pushed to the edge of the cell.*

Right *Colored scanning electron micrograph of fat (adipose) tissue. Individual adipocytes are held together with connective tissue, mostly collagen.*

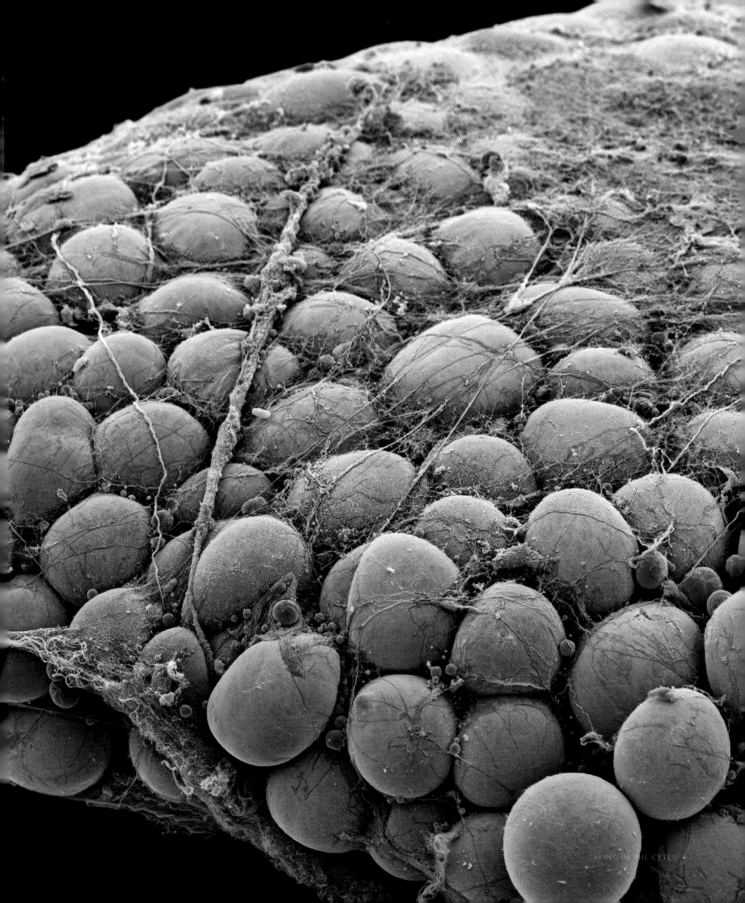

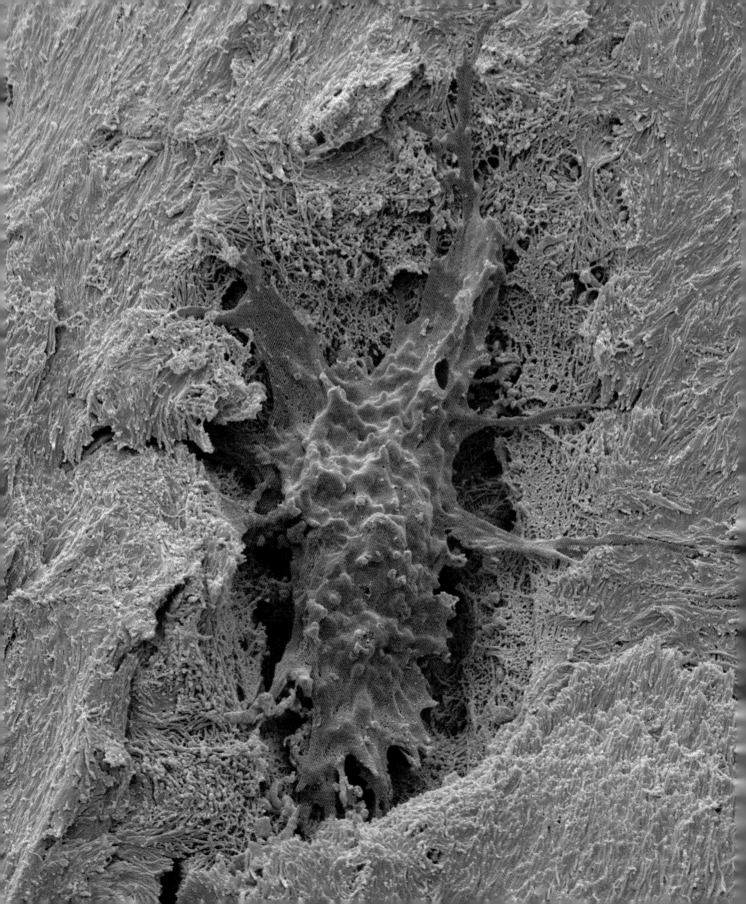

Left Colored scanning electron micrograph of an osteocyte (red) in a lacuna (depression) in mineralized osteoid (gray).

Osteocytes (bone cells)

Bone is a dynamic material that is constantly being remodeled: cells called osteoblasts build it up, and cells called osteoclasts reabsorb it. Both cell types move around inside the bone in an amoeboid fashion. To make new bone, osteoblasts lay down tightly woven collagen fibers with a few other proteins mixed in—a composite material called osteoid. Crystals of carbonated hydroxyapatite (a form of calcium phosphate) mineralize the osteoid. Osteocytes are osteoblasts that become trapped within the matrix of mineralized osteoid; they are the most common type of cell in mature bone.

Osteocytes can live for decades. Once they are in position, they are responsible for sensing stresses on the bone. They feed this information to each other via long dendrites that interconnect them with their neighbors. The signals they share help to make sure the bone remodeling is appropriate to the level of stresses on the bone. They also produce growth factors if a bone sustains damage, encouraging the rapid remodeling necessary to make the bone strong again.

Above Light micrograph of an osteon—the basic unit of compact bone. At the center is the Haversian canal, which carries nerve and blood supplies. The osteocytes are the dark shapes and the thin lines that connect the osteocytes are microscopic canals, or canaliculi.

Right Collagen stains pink in this light micrograph of osteocytes in growing bone.

TAKING IN THE CYTES

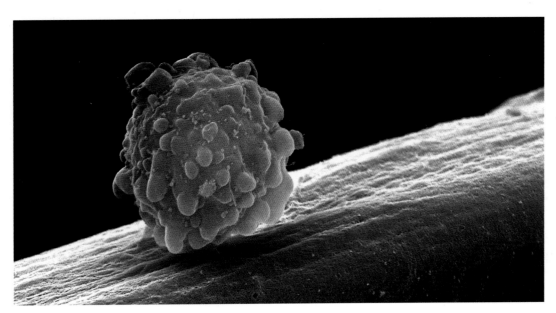

Left *Colored scanning electron micrograph of a multipotent muscle stem cell migrating along the polysaccharide surface of a muscle fiber.*

Below left *Illustration of a sarcomere, the basic unit of striated muscles. The Z line boundary is what gives this kind of muscle its striped appearance under a microscope.*

Myocytes (muscle cells)

There are two main types of muscle cell: smooth and striated (meaning it is patterned with fine parallel lines). Both work in the same way, with protein fibers contracting and relaxing to change the cell's shape, although the inner structure of the two types is slightly different. Smooth muscle, which is involuntary (not under conscious control), is responsible for important housekeeping tasks, such as squeezing food through the digestive system and opening and closing our iris as light levels change. Striated muscle comes in two varieties. Involuntary cardiac striated muscle keeps our hearts beating without us having to think about it, while voluntary skeletal striated muscle makes our bones move relative to each other and keeps our lungs breathing.

Striated muscle cells are more interesting and more beautiful to look at than smooth muscle cells—and most interesting among them are the skeletal cells. These are long, tubelike structures that contain many nuclei, because they are formed from several cells that have merged together. The cell membrane is a phospholipid bilayer, as with any other cells, but it also has a thin layer made of the sugary polymer polysaccharide tightly bound to it. The resulting cell coating is called the sarcolemma. Beneath that outer coating, the inside of the cell is dominated by thick bundles of fibrous protein called myofibrils. Clustered around these structures, perhaps not surprisingly, are many mitochondria, which can provide a constant supply of ATP to enable the myofibrils to do their job.

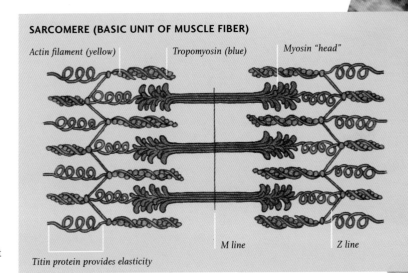

SARCOMERE (BASIC UNIT OF MUSCLE FIBER)

Actin filament (yellow)   Tropomyosin (blue)   Myosin "head"

Titin protein provides elasticity   M line   Z line

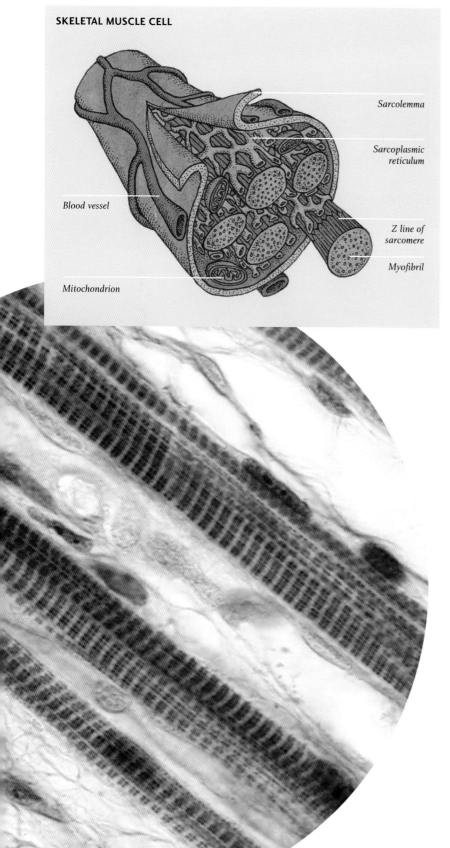

SKELETAL MUSCLE CELL

- Sarcolemma
- Sarcoplasmic reticulum
- Blood vessel
- Z line of sarcomere
- Myofibril
- Mitochondrion

The job of a myofibril is to contract and relax. When it is relaxed, it can be pulled back to its original length by other muscles (muscles are arranged in opposing pairs—think biceps and triceps). Each myofibril is divided along its length into sections called sarcomeres. The contraction of the myofibril is the result of the sarcomeres shrinking, as two fibrous proteins, actin and myosin, slide past each other. The actin forms thin strands that are connected to both ends of the sarcomere—half of the strands to one end and half to the other. The myosin strands are thicker and are interspersed among the actin strands. When ATP molecules, produced by the mitochondria bind to the myosin the ends of the myosin molecules attach to specific sites on the actin strands. They pull the actin, drawing the ends of the sarcomere together by a tiny amount. Then the myosin lets go of the actin, ready to attach and pull again a tiny fraction of a second later. Each power stroke requires another ATP molecule.

When the muscle is relaxed, a protein called tropomyosin covers over the parts of the actin molecules where the myosin attaches, so the muscle remains inactive. When a nerve impulse arrives at the muscle cell, it causes calcium ions to be released into the cytoplasm. These attach to proteins along the length of the tropomyosin, which then changes shape to reveal the sites where myosin can attach, enabling the sarcomere to contract. It is truly remarkable to think that every move we make is the result of tiny straggly ends of myosin molecules grabbing hold of and pulling actin fibers, activated by a burst of calcium ions.

Left Light micrograph of striated skeletal muscle fibers. Every muscle contraction you have ever made was the result of motor proteins "walking" up and down inside tiny subunits in these fibers.

Above left Illustration of a skeletal muscle cell, showing the bundles of myofibrils, each one composed of a long line of successive sarcomeres.

TAKING IN THE CYTES

Retinocytes (rod and cone cells)

The term retinocytes, rarely used now, refers to all the cells of the retina, the layered structure at the back of the eye. There are several types of retinocytes but the most interesting and important are the ones that enable humans to see: the rod and cone cells. These have a very similar structure but slightly different purposes. The graph here shows the sensitivity of rod and cone cells to light entering the retina. The latter—of which there are three different types, each sensitive to a specific range of frequencies of light, roughly equivalent to red, green, and blue—are not particularly sensitive, so they do not work in low light, but they are responsible for color vision. Rod cells are extremely sensitive to light—far more so than cone cells—and are responsible for vision in low light. But they only detect light and dark, so they only enable black-and-white vision with a peak in the green part of the spectrum. Likewise color sensitivity peaks in that same region of the spectrum. Overall rod cells outnumber cone cells by about 20 to 1; a typical retina, with an area of 1 1/2 square inches (10 square centimeters), has around 120 million rod cells and about six million cones. The cones far outnumber the rods, however, in the part of the retina that gives the sharpest vision, called the fovea.

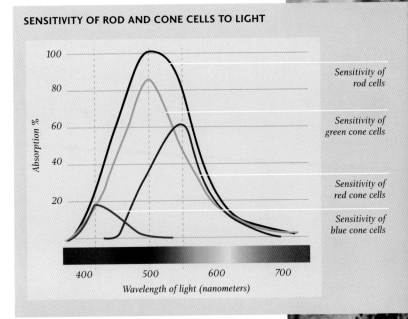

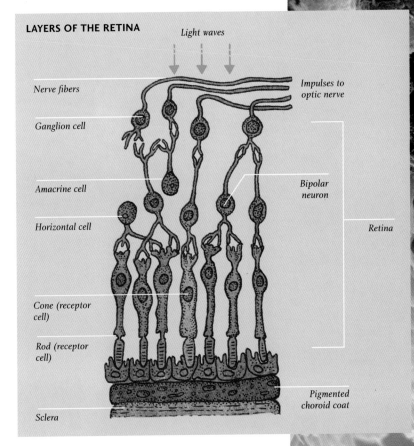

Facing page *Colored scanning electron micrograph of a section through a human retina. Light enters from the top and passes through several layers of cells. The rods and cones are the white and yellow cells respectively, at the bottom of the image.*

TAKING IN THE CYTES

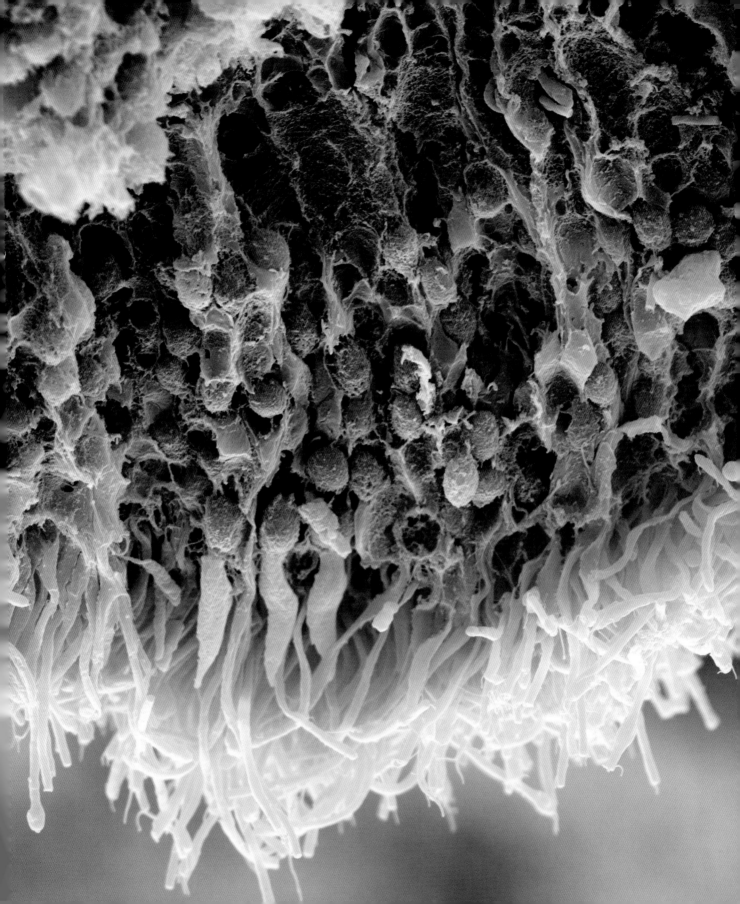

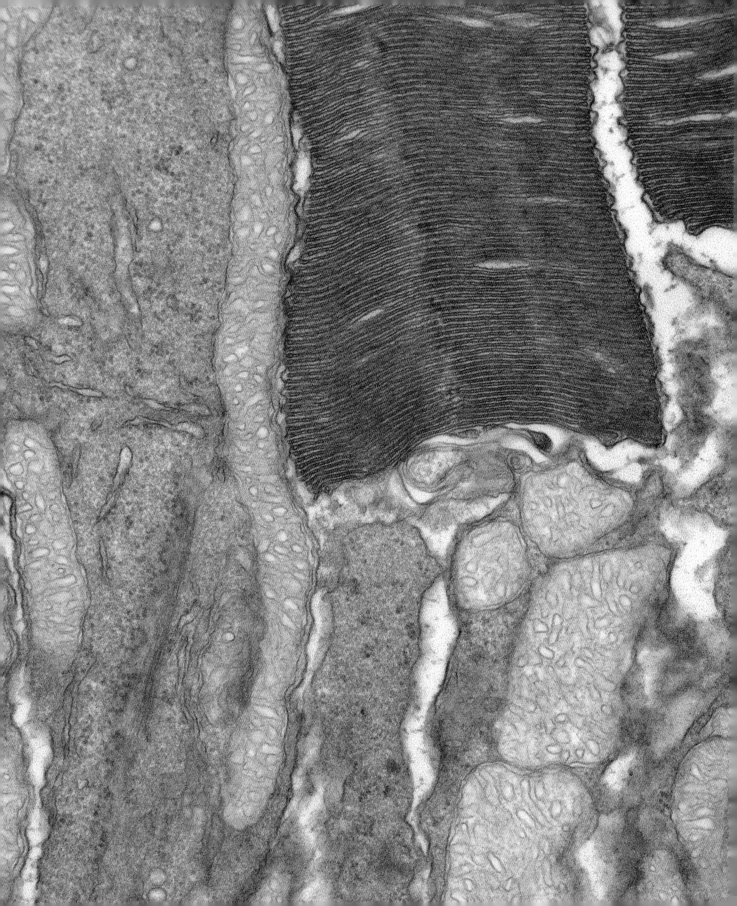

Left *Colored transmission electron micrograph of a section through a rod cell showing the inner segment (blue) filled with mitochondria (green) and outer segment made up of stacks of folded membranes (brown).*

Below *Illustration of a rod cell. The part that links the two parts of the cell is a fixed cilium. The axonal projection connects to cells in other layers of the retina that pass signals on, ultimately to cells in the optic nerve.*

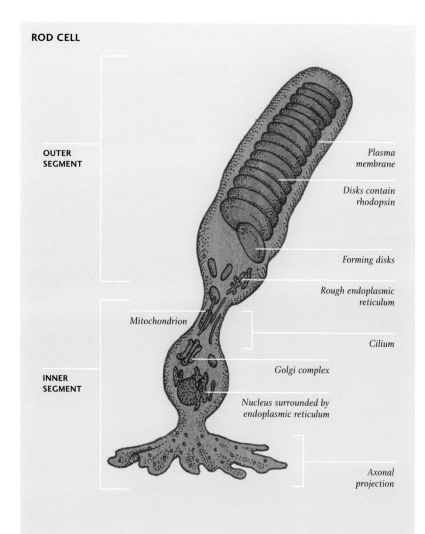

Rod and cone cells have two very distinct ends. In both types, one end of the cell contains the nucleus and most of the organelles. The other end is dominated by stacks of membranes (actually modified cilia), formed by the cell membrane folding in on itself. In rod cells, this end is rod-shape and in cone cells, it is cone-shape. Each type of cell constantly pumps sodium ions out at one end and takes them back in at the other. Because sodium ions are electrically charged, this circulation constitutes a tiny electric current called the dark current. It is the interruption of this current, when the cell absorbs light, which creates the signal that is ultimately passed along the optic nerve to the brain.

Embedded within the membrane stacks at the business end of each rod and cone cell are pigment molecules. It is these that actually absorb incoming light. The pigment in rod cells is rhodopsin and it is sensitive to light across a wide range of frequencies, centered on blue-green. The equivalent pigments in cone cells are called photopsins. Both rhodopsin and photopsin molecules are in two detachable parts: one large opsin and a smaller retinal molecule (retinal is made from vitamin A). When a pigment molecule absorbs light, the retinal molecule changes shape and temporarily breaks away. This interferes with the circulation of sodium ions and begins the process of creating a nerve signal.

Surprisingly, perhaps, the light-sensitive end of each rod and cone cell is found in the deepest layer of the retina, farthest away from the front of the eye. Light has to pass through eight layers of cells and two membranes before it reaches the stacks containing the pigments. At the end of the cell farthest away from the light-sensitive part, and closest to the front of the retina, the cell passes on signals to nerve cells; these are bundled together to form the optic nerve that connects to the brain.

Neurocytes (neurons)

An average human brain contains around 85 billion neurocytes—nerve cells, or neurons. Each neuron in the brain receives signals from up to a thousand other neurons and, depending upon the overall effect of these signals, either fires or doesn't fire, passing on its own impulse to many others. So the brain contains an incredibly complex network of interconnected neurons, all busily communicating with each other. Those interconnections form and break as we experience the world and learn new things. Somehow—no one exactly knows exactly how—the signals zipping here and there in the brain give us awareness, recall, and the ability to act, as well as subconsciously monitoring and regulating the vital organs of the body.

All this activity would be for nothing were it not for neurons that connect the brain to the rest of the body and to the outside world. Signals enter the brain via sensory, or afferent, neurons, whose origins are in the sensory receptors, such as rod and cone cells in the retina and pain and pressure receptors in the skin. Signals leave the brain via motor, or efferent, neurons and cause muscle cells to contract. Neurons come in several different shapes and sizes, but they all share certain features: they gather input signals at extensions called dendrites, they process the signals in the cell body, or soma, and they produce an output signal that passes along an extension called an axon. The end of a typical axon has many branches, called terminals, at which other neurons connect their dendrites—or, in the case of motor neurons, at which a muscle cell receives signals.

Nerve impulses are electrical in nature. When it is not firing, every neuron has more positive charge on the outside than the inside, thanks to a remarkable cluster of molecules that sits in the cell membrane. It is called the sodium-potassium pump, and it pushes three sodium ions (3Na+) out of the cell for every two potassium ions (2K+) that it pulls into the cell. The imbalance of charge across the membrane gives the inside of the cell a voltage of about minus 70 millivolts relative to the outside (the inside is

Right *Colored scanning electron micrograph of neurones from the brain. Cell bodies are brown, and the axons and dendrites are clearly visible.*

Below *Illustration of the sodium-potassium pump, which creates the resting potential by pumping more sodium ions (Na+) out of the cell than it pumps potassium ions (K+) into the cell. The resulting resting potential leaves the interior of the cell more negatively charged than the outside, as shown in the final step.*

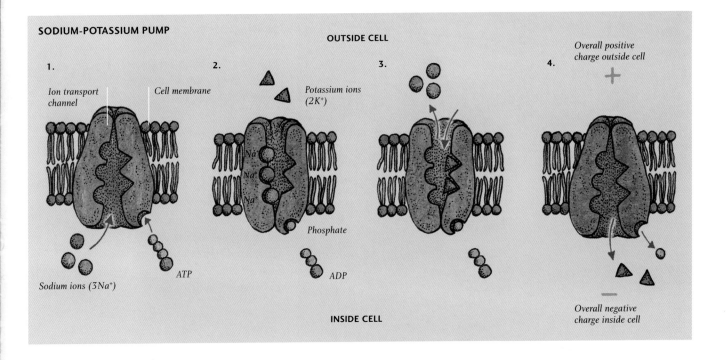

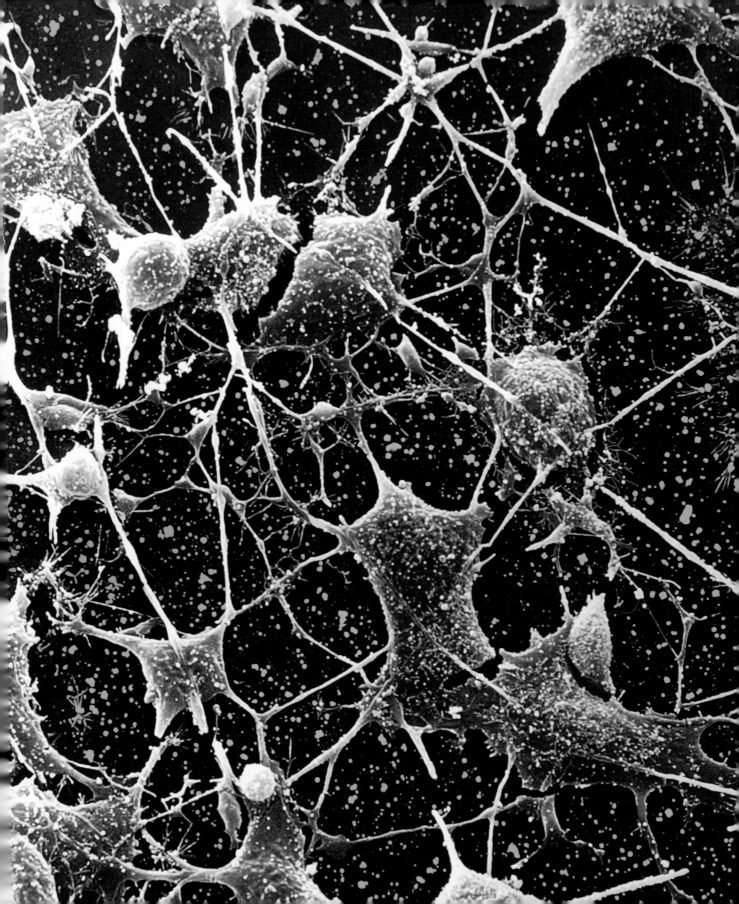

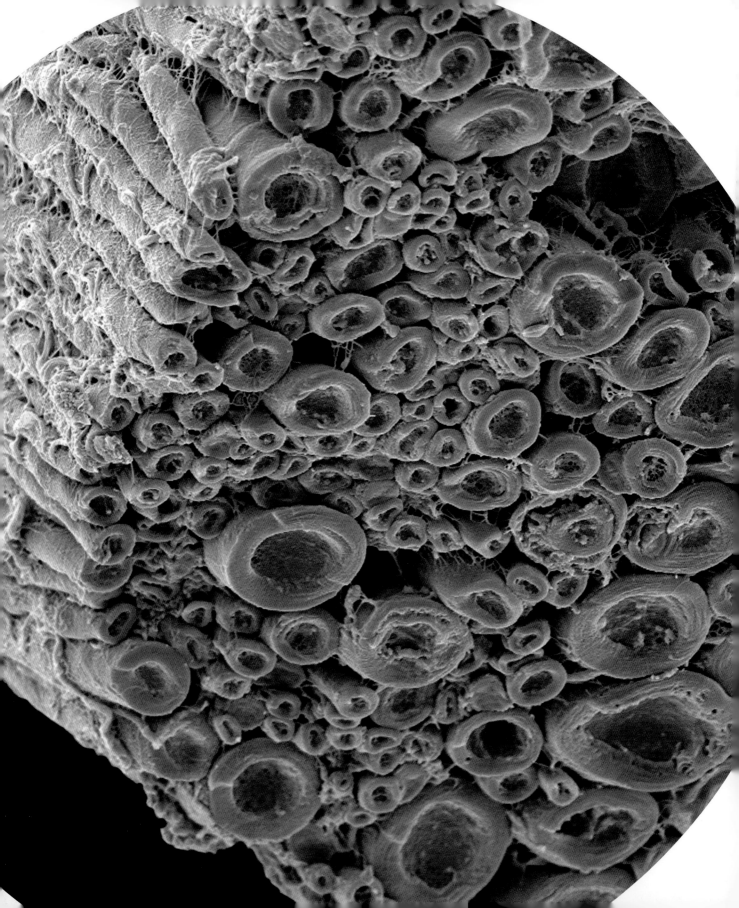

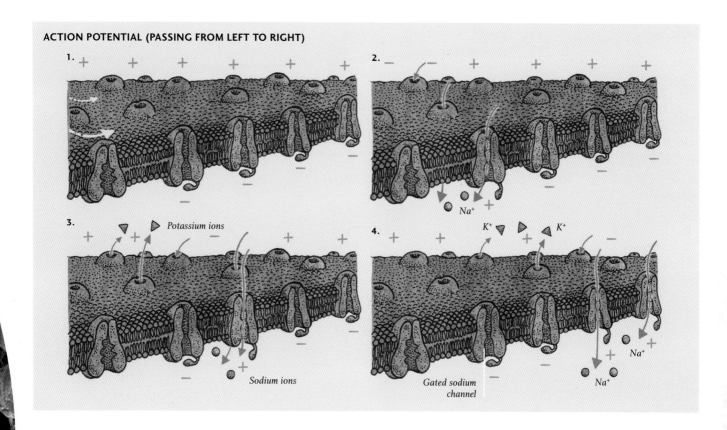

ACTION POTENTIAL (PASSING FROM LEFT TO RIGHT)

Above This sequence shows how a nerve signal moves along a neuron. When a neuron fires, the resting potential (1) is disturbed and sodium ions are pulled into the cell through gated channels (2). The channels close afterward (3), ensuring that the signal passes only one way (4).

Left Colored scanning electron micrograph of a freeze-fractured section through a bundle of myelinated nerve fibers. Myelin sheaths (purple) can be seen surrounding the axons (brown). Connective tissue (blue) is also visible.

more negatively charged than the outside). This voltage is called the resting potential. The cell remains in this polarized state until a dendrite receives a signal from another neuron, or the cell decides to fire. In either case, a disturbance in the ion concentration, called the action potential, passes along the axon or dendrites.

The action potential moves along the neuron's axon and dendrites because of the presence of protein clusters called gated sodium channels, which are dotted all over the membrane. These channels are normally closed, but when an action potential arises, they open and enable sodium ions to rush into the cell, depolarizing that part of the membrane. The voltage across that part of the membrane initially becomes zero and then for a moment rises above zero—and that causes adjacent sodium channels to open.

It is important that nerve impulses travel in one direction only—from input at the dendrites to the cell body, and from the cell body to the output at the axon terminals, and from brain to muscle, and from receptor to brain. The traveling action potential can go only one way, because the channels close shortly after the potential reaches its maximum and remain closed for a few milliseconds. These cannot, therefore, be opened again by the action potential they have just passed on to adjacent channels down the line. This remarkable process is how nerve signals pass within the brain and throughout the body. Some neurons are sheathed by insulated sections of a fatty material called myelin. These enable the action potential to jump quickly from gated sodium channels in one uninsulated node to the channels in the next. Neurons that are myelinated in this way transmit impulses much faster than unmyelinated neurons.

TAKING IN THE CYTES

The transmission of a nerve impulse from one neuron to another is equally remarkable. The neurons are not physically connected; instead, there is a small gap called a synapse (see chapter five) between the axon terminal of one neuron and the dendrite of another. The nerve signal is transmitted across the synapse by the release of compounds called neurotransmitters from the axon terminal. Neurotransmitters are manufactured in the rough endoplasmic reticulum just outside the nucleus in the cell body. The Golgi apparatus nearby (see chapter two) squirrels away the neurotransmitter molecules into vesicles, bound for the neuron's cell membrane. The molecules are then "walked" along the cytoskeleton by motor proteins to the axon terminals. Here, they stay, ready to be released quickly, as the vesicles burst like bubbles at the membrane in the axon terminal.

Neurotransmitters pass across the synapse and dock into receptor proteins in the membrane of the dendrite on the other neuron. Some receptors cause the membrane to remain polarized, in which case they inhibit the receiving neuron's membrane from creating an action potential. Other receptors will depolarize the membrane, exciting the membrane to create an action potential that travels along the dendrite toward the cell body. Meanwhile, neurotransmitter molecules are released back into the synaptic gap and are reabsorbed by the first neuron, ready for the next time it fires.

A neuron has many dendrites, each receiving signals from another neuron. If enough dendrites are excited and create action potentials at the same time, reaching a threshold, they create an output signal at the other side of the cell body, at the very beginning of the axon—a part of the cell known as the axon hillock. When that is achieved, the neuron will fire.

Of all the cells in the body, neurons are perhaps the most amazing: they really do make you think.

Right *Colored scanning electron micrograph showing synapses between axons (purple) and a nerve cell grown in culture (yellow). The axons do not actually touch the cell body; instead, neurotransmitters pass across a small gap.*

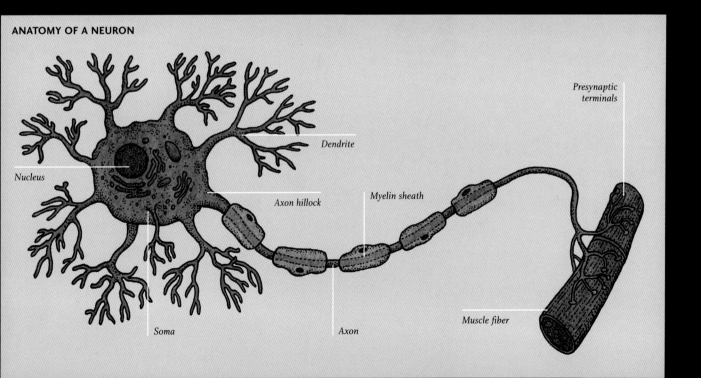

ANATOMY OF A NEURON

Nucleus · Dendrite · Axon hillock · Myelin sheath · Presynaptic terminals · Muscle fiber · Soma · Axon

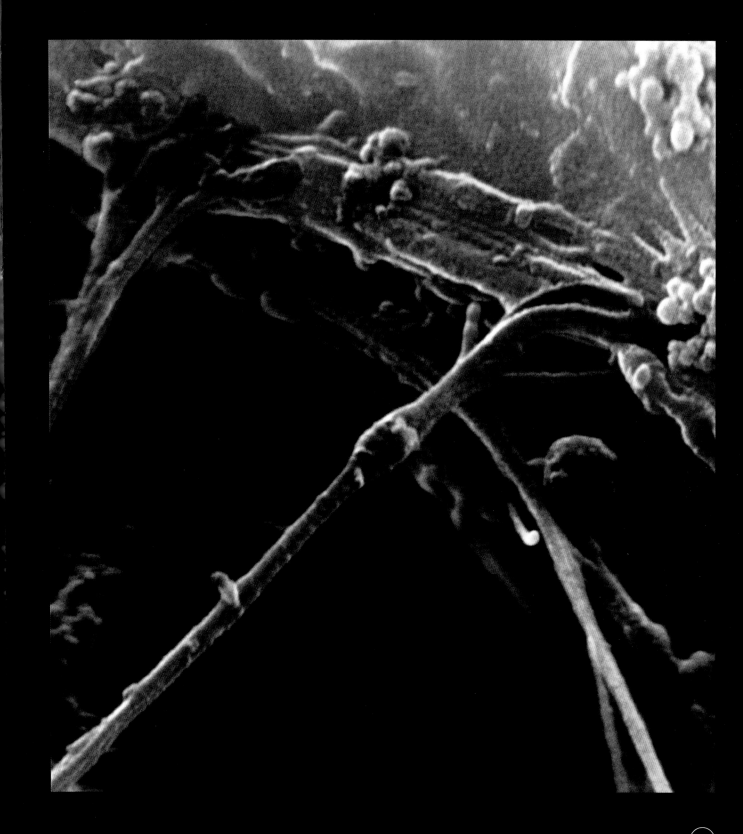

TAKING IN THE CYTES

Glossary

Aerobic respiration
Metabolic process in some cells, in which ATP is produced by the oxidation of carbohydrates (respiration) involving oxygen.

Amino acid
A small molecule consisting of carbon, hydrogen, oxygen and an amine group (NH2)—some also contain sulfur. Amino acids join together to form proteins.

Anaerobic respiration
Metabolic process in which ATP is produced by the oxidation of carbohydrates (respiration) not involving oxygen.

Apoptosis
Controlled cell death, after irreparable damage to DNA or the infection of a cell by viruses or bacteria.

Archaea
Together with Bacteria and Eukaryotes, one of the three domains of living organisms.

ATP (Adenosine triphosphate)
An energy-rich molecule that enables most molecular-scale cellular processes. As it supplies energy, it breaks down to form adenosine diphosphate (ADP) and a free phosphate group.

Budding
A form of asexual reproduction or cell division that results in two identical daughter cells. A new cell grows as a "bud" on the existing cell, unlike simple cell division, in which one cell divides in two.

Cell cycle
A repeating series of events by which a cell replicates its contents—in particular the contents of its nucleus by mitosis—then divides.

Chromatin
Stringy material found in the nucleus of eukaryotic cells and composed of DNA packed around histone proteins.

Chromosome
Structures in eukaryotic cells, made of chromatin in its most compacted state. Chromosomes form during mitosis. Chromosome is also the name given to any isolated length of DNA, so the DNA inside a prokaryotic cell is sometimes referred to as a chromosome.

Codon
A set of three bases (nucleotides) along a length of DNA. In general, each codon codes for a specific amino acid and is important in the process of translation, by which the protein recipes in genes make protein molecules.

Cytoskeleton
A dynamic network of protein filaments and tubes found inside eukaryotic cells. It is responsible for a number of vital processes, including transporting vesicles around the cell and dragging duplicated chromosomes apart during mitosis.

Diploid
Describes a (eukaryotic) cell that has two sets of corresponding chromosomes. Most cells of an organism are diploid. *See* Haploid.

DNA (Deoxyribonucleic acid)
The double helical molecule that carries genetic information in all living organisms.

Endoplasmic reticulum
An organelle found in eukaryotic cells.

Eukaryotes
Together with Bacteria and Archaea, one of the three domains of living organisms. Eukaryotic cells are larger and more complex than the prokaryotic cells of the other two domains.

Gamete
A haploid (eukarytotic) cell, produced by meiosis, which can join with another gamete to form a new, unique organism.

Gametophyte
A part of a multicellular (eukaryotic) organism in which all the cells are haploid gametes. All plants and fungi produce gametophytes for at least some of their life cycles.

Golgi complex
An organelle found in eukaryotic cells, made of folded phospholipid membranes, upon which protein molecules that have been made by transcription may be altered by adding other proteins or removing sections or by folding.

Gram stain
A chemical stain that is routinely used in the identification of bacteria. Those bacteria that have an outer membrane as well as an ordinary cell membrane have thinner cell walls, so do not take up the stain and are "gram negative."

Haploid
Describes a (eukaryotic) cell that has only one set of each chromosome. Sex cells (gametes) are haploid. *See* Diploid.

Meiosis
A process in eukaryotic cells consisting of two rounds of cell division, the result of which is four haploid gamete cells, or "sex cells."

Micron
One millionth of a meter, equal to one thousandth of a millimeter (0.001 mm), about 0.00004 inches.

Mitochondrion
An organelle in eukaryotic cells that are described as cellular power stations. They are composed of phospholipid membranes, on which molecules of ATP synthase produce ATP.

Mitosis
A crucial part of the cell cycle in eukaryotic cells, in which already-duplicated chromsomes are pulled apart into two new nuclei, ready for cell division to take place.

mRNA see RNA

Multipotent see Potency

Nucleic acid
Large (long) molecules found in living organisms, made up of smaller molecules (nucleotides). The two most important nucleic acids are DNA and RNA.

Nucleobase
A nitrogen-containing compound found in nucleotides and, therefore, crucial parts of nucleic acids. Four nucleobases are found in DNA: adenine (A), cytosine (C), guanine (G), and thymine (T). The same is true for RNA, except that thymine is replaced by uracil (U).

Nucleotide
A compound whose molecules bond together to form nucleic acids. A nucleotide molecule is composed of a nucleobase plus a sugar and a phosphate group. Four nucleotides make up DNA, and are distinguished by the nucleobase they contain (A, C, G, and T).

Organelle
A structure with a specific function found inside cells. Only eukaryotic cells have organelles.

Oxidation
The opposite of Reduction. The part of a chemical reaction in which atoms or molecules lose electrons. Oxygen has a high affinity for electrons, and reactions with oxygen are archetypal example of oxidation.

Pathogen
Any organism that does harm to another. Many bacteria are pathogenic to humans but most are not.

Peptidoglycan
A compound composed of sugars and amino acids that forms the cell wall of bacteria.

Phospholipid
A compound whose molecules have a hydrophilic end, which will mix with water, and a hydrophobic end, which will not. Phospolipids form double layered membranes around cells, around organelles, and around vesicles.

Pluripotent see Potency

Potency
The extent to which the daughters of a cell in a multicellular organism can become any of a range of different types of cell. The daughters of multipotent cells can become any of several types within a certain tissue, while the daughters of pluripotent cells can become almost any type of cell. Embryonic stem cells are totiptent; through repeated cell divisions, they can give rise to an entire organism.

Prokaryote
A single-celled organism that is not a eukaryote—in other words, an archaeon or a bacterium. Prokaryotic cells are simpler than eukaryotic ones; they have no nucleus or organelles.

Reduction
The opposite of Oxidation. The part of a chemical reaction in which atoms or molecules gain electrons.

Ribosome
A molecular-scale "machine" that provides a site at which amino acids carried by tRNA are built up to form proteins, according to the instructions carried on lengths of mRNA. *See also* RNA.

RNA
A nucleic acid that is single stranded. Inside cells, lengths of messenger RNA (mRNA) are made as copies of genes (during transcription) and short lengths of transfer RNA (tRNA) carry amino acids, ready to be joined together to make proteins (during translation).

Stem cell
A pluripotent, multipotent, or totipotent cell in a multicellular organism. *See* Potency.

Totipotent see Potency

Transcription
The process by which genes are copied into lengths of mRNA. *See* RNA.

Translation
The process by which proteins are built up inside ribosomes. *See* RNA.

tRNA see RNA

Vesicle
A membrane-enclosed globule in which proteins and other molecules are transported or stored inside (eukaryotic) cells.

Zygote
A diploid cell produced by the union of two (haploid) gametes.

Index

A
Abbe, Ernst 18, 26
acetic acid 152
adenosine diphosphate (ADP) 52, 54
adenosine triphosphate (ATP) 25, 51–2, 54, 55–6, 81, 176–7
adenosine triphosphate (ATP) synthase 25, 54, 55–6
adipocytes 172–3
alcohol 55, 150
algae 8, 32–4, 56, 93, 126, 129–31, 147
alleles 64
alternation of generations 77
amino acids 25, 38, 45, 47
anaerobic respiration 34, 55
anaphase 66, 67, 69
animalcules 14
antibiotics 32
  resistance 106, 125
antibodies 156
antigens 156
antiseptic 153
antithrombin 87
Apicomplexa 115
apoptosis 157–60, 162, 163, 172
Archaea 30, 32, 40, 42, 51, 55, 61, 92–7, 100–3, 108
  see also prokaryotes
artificial selection 80, 84
asexual reproduction 60, 70–1, 77, 88, 109
autotrophs 56, 146–7

B
bacteria 14, 19, 22, 30, 32, 50–1, 55, 61, 89, 92–5, 100–8
  antibiotics 32, 106, 125
  biofilms 124–5
  cell death 148–56
  classification 98–9
  colonies 124–7
  cyanobacteria 100, 124, 126–7, 132–3
  genome 40, 82
  nitrogen fixation 101–2, 132–3
  phages 145
  platelets 169
  spores 98, 147–9, 150–2
  thermophilic 96
  transgenic 86–7
  see also prokaryotes
bacteriophages 145
binary fission 16–17, 61
biofilms 124–5
blastula 135
blood cells 12, 22, 78, 106, 114, 133, 162, 166–9

bone cells 160, 175
Borlaug, Norman 84
Boveri, Theodor 21
Brown, Robert 16
bubonic plague 105
budding 8, 70, 71
Buffon, Comte de 16

C
Calvin cycle 56
Cambrian Explosion 130
cancer 78, 92, 161–3
capillaries 12
carbon cycle 100
carrier proteins 38
cartilage 121
caspases 157
cell cycle 60–9, 160, 163
cell death 9, 133, 140, 142–63
cell division 18–19, 59–77, 134–5
cell membrane 34–7, 55, 61, 95, 122, 161
cellular respiration 25, 34, 36, 51, 55
cellulose 21, 33, 56, 100, 121–2
centrosomes 66–7
chemotherapy 163
chitin 33, 122
chlorophyll 56
chloroplasts 21, 30, 56–7, 95
cholera 105
cholesterol 34
chromatids 65–7, 71
chromatin 18, 19, 21, 136–7, 140
chromosomes 18, 19, 20–1, 40, 64–73, 89, 161
cilia 50–1, 110, 155, 178
ciliates 110, 113
clones 60, 88–9, 135
coccolithophores 113
codominance 64
collagen 21, 121, 172
colonies 124–7, 130, 132–3
complement proteins 155, 156
contact inhibition 162
Crick, Francis 23
cyanobacteria 100, 124, 126–7, 132–3
cytoblastema 16, 17, 18
cytokinesis 63, 73
cytoplasm 33, 40, 46, 50, 61, 117, 169
cytoskeleton 48–51, 117, 122
cytosome 110
cytotoxic T cells 157

D
dark field microscopy 26–7
Darwin, Charles 80–1
dendrites 171, 175, 182, 185, 186
dendritic cells 155–6, 157
deoxyribonucleic acid (DNA) 22–5, 27, 30, 38–45, 55–6, 84–5
  binary fission 61
  cell cycle 64–9
  cell death 157, 158
  E. coli 106
  synthesizers 89
  variation 71, 78–83
deoxyribonucleic acid (DNA) methyltransferase 137
desmosomes 49, 122
diatomaceous earth 113
diatoms 113
differentiation 60, 132–4, 136, 160
diffusion 36, 55, 109
disease
  bacteria 104–7
  cancer 78, 161–3
  protists 114–15
Dolly the sheep 88, 135
dominance 20–1, 79
double fertilization 77
Dumortier, Barthélemy 17
Dutrochet, Henri 16
dyneins 31, 50–1

E
eggs 20, 70, 74–5, 88, 136
elastin 21, 121
electron microscopes 26
electron transport chain 54
embryos 17, 87, 134–5, 138–9, 140
endocytosis 37
endoplasmic reticulum 19, 30, 46, 186
endosymbiosis 56, 95
entropy 144
enzymes 21, 25, 38, 137
erythrocytes 166
Escherichia coli 32, 61, 106, 150
eukaryotes 30–1, 40, 42, 55, 93
  cell cycle 60, 62–9
  cytoskeleton 48–9
  unicellular 108–11
evolution 24, 25, 71, 79–84, 92, 106, 130, 149
extracellular matrix 33, 121–2
extremophiles 94, 96–7
eyes 178–81

F
fat cells 172–3
fermentation 52, 55

fertilization 20
fibroblasts 121
filamentous bacteria 101
fimbriae 32
flagella 31, 51, 110
flagellates 110, 113
Flemming, Walter 18, 19
fluorescence microscopy 27
food poisoning 106, 150
Franklin, Rosalind 23
frustrules 113
fungi 32–3, 129–30, 147
  ant farming 102–3
  reproduction 77

G
gametes 20, 70–7, 80, 114
  see also eggs; sperm cells
gametophyte 77
gap junctions 122
gatekeeper proteins 44
gene expression 136–7
genes 21, 23, 39
genetically modified (GM) crops 87
genetic engineering 84–7, 106
genetics 20–4, 38–45, 78
  see also deoxyribonucleic acid (DNA); reproduction
genomes 24, 40, 70, 81–2, 86–7, 140
  bacteria 40, 82
  synthetic 89
giardiasis 115
goats, transgenic 87
Golgi apparatus 19, 30, 47, 186
gram-negative/-positive bacteria 32
Green Revolution 84
grex 117
Griffiths, Frederick 22
growth 8
growth factors 160–1, 162, 163

H
Haeckel, Ernst 108–9, 110, 116, 138–9
Halococcus salifodinae 96
Harvey, William 12
Hayflick, Leonard 161
heat shock proteins 148, 152
HEPA filters 153
heterocysts 132–3
heterotrophs 146–7
histones 42–3, 95, 137
Hooke, Robert 13, 33
horizontal gene transfer 79, 80
hox genes 140
Human Microbiome Project 92

I
immortal cells 160–3
immune system 155–7
insulin 37, 38, 47, 86
intermediate filaments 48, 49

K
keratinocytes 171
kinesins 31, 50–1
Knoll, Max 26
Kossel, Albrecht 22
Krebs' cycle 54

L
lactic acid 55
Leeuwenhoek, Antony van 14
leptin 172
Levene, Phoebus 22
ligands 36–7
lignin 33, 122
Lister, Joseph 153
Lister, Joseph Jackson 18
Lohmann, Karl 25
loricae 113
Lovelock, James 144
luminescent bacteria 102
lymphocytes 156, 157
lysosomes 47, 155
lysozyme 150, 154–5

M
macrophages 155–6, 157
malaria 114, 117
Malpighi, Marcello 12
meiosis 20, 71–7, 114, 117
melanin 170–1
melanocytes 170–1
Mendel, Gregor 20, 79
Merismopedia 126
meristematic cells 134
messenger RNA (mRNA) 44–5, 46
metaphase 66, 67, 69, 72
methanogens 100
Methanosarcina 96, 127
Meyen, Franz 16
MHC class I complex 157
MHC class II complex 156–7
microfilaments 48–9
microRNAs 140
microscopes 12–15, 18, 26–7
microtubules 48, 50–1, 62–3, 67–9, 72, 117, 163
Miescher, Friedrich 22
mitochondria 19, 25, 30, 55–6, 95
mitosis 18–19, 63, 66–9, 77, 117, 134, 136, 160
Möbius, Karl 19
Morgan, Thomas Hunt 21
motor proteins 31, 50–1
MRSA 32, 106
multicellularity 118–41
muscle 49, 122, 176–7

mutations 78, 80, 92, 158
myelin sheath 185
myocytes 176–7
myofibrils 176–7

N
natural selection 80–1
  see also evolution
necrosis 158
nerve cells 182–6
neurocytes 182–6
neurotransmitters 47, 122, 186
neutrophils 155–6
Newton, Isaac 16
nicotinamide adenine dinucleotide (NADH) 54
nitrogen fixation 101–2, 132–3
Nostoc 127, 132
nucleic acids 22–5, 81
  viruses 145
  see also deoxyribonucleic acid (DNA); ribonucleic acid (RNA)
nucleosomes 137
nucleotides 39–40, 44–5, 64–5, 78, 82, 85, 89, 140
nucleus 16, 18, 20, 21, 30, 40, 46
  protists 110
  transfer 88

O
oomycete 116
organelles 18, 19, 30, 50, 55–6
origins of life 24, 25
osteocytes 175
oxidation 52, 54
oxidative phosphorylation 54

P
Palade, George 24
parasites 93, 114–15
pasteurization 152
Pasteur, Louis 152, 153
pectin 33
phagocytes 155–6
phase contrast microscopy 27
phospholipids 34, 36, 46, 145
photosynthesis 25, 56, 100, 132–3
picophytoplankton 109
plants 33, 47, 56–7, 115, 121–2, 129–31
  cell death 133, 157–8
  meristematic cells 134
  multicellularity 120–2
  reproduction 60, 77
  vascularization 131
plasmids 79
plasmodesmata 122
Plasmodium 114–15, 117
plasmodium 117
platelets 169
pluripotent cells 134, 135, 140
polar bodies 75
polysaccharides 32, 121, 176
post-mitotic cells 160

potato blight 115
precursor cells 133, 134
prokaryotes 30–1, 40, 56, 60–1, 79
  single cell 92–107
  see also Archaea; bacteria
prometaphase 66, 67, 68
prophase 66, 68, 72
proteins 21–2, 24, 38–47, 121
  cell death 157
  cytoskeleton 48–9
  "p53" 158, 162
protists 108–15, 130
protophyta 108
protoplast 33
protozoa 108, 110
Pyrococcus furiosus 97

Q
quorum sensing 125

R
radiotherapy 163
receptor proteins 36, 38
recessive genes 20–1, 79
red blood cells 12
reduction 54
Remak, Robert 17
reproduction 8, 16–17, 58–89
  asexual 60, 70–1, 77, 88, 109
  sexual 70–9
  vegetative 60, 87
respiration 52, 54–5
  anaerobic 34, 55
  cellular 25, 34, 36, 51, 55
restriction enzymes 85
retinocytes 178–81
ribonucleic acid (RNA) 22, 24–5, 30–1, 44, 137, 140
ribonucleic acid (RNA) polymerase 44
ribosomes 24, 31, 45–7, 55–6, 61
rod and cone cells 178–81
rough endoplasmic reticulum (RER) 46, 137
Ruska, Ernst 26

S
Salmonella 32, 150
Schleiden, Matthias 16
Schrödinger, Erwin 144
Schwann, Theodor 16, 17, 18
secretory vesicles 37, 46
self-destruction 9
sex cells *see* gametes
sexual reproduction 70–9
sickle cell disease 78
signaling proteins 38–9
single-celled organisms 90–117
single-nucleotide polymorphism (SNP) 78
size of cells 8, 109–10
slime molds 116–17
smooth endoplasmic reticulum (SER) 46–7

sodium channels 185
sodium-potassium pump 182
somatic cells 78, 88, 135
sperm cells 20, 74, 77
spores 77, 98, 114, 116, 117, 147–9, 150–2
sporophytes 77
staining 18, 98
stem cells 133–5, 140, 161, 162
sterilization 152–3
Sulfolobus 97
Sumner, James 21
Sutton, Walter 21
symbiosis 55–6, 93, 102
synapse 122, 186
synthetic biology 88–9
Syringammina fragilissima 109, 110

T
Tait, Lawson 153
telomerase 162
telomeres 161–2
telophase 66, 67, 69, 72–3
thrombocytes 169
totipotent cells 134, 136
transcription 44, 95
transcription factors 137, 140
transfer proteins 36–7
transfer RNA (tRNA) 45
translation 45, 95
transposons 79, 80
tropomyosin 177
tubulin 50–1

U
uracil 44

V
vacuoles 47, 110
variation 71, 78–83, 106
vegetative reproduction 60, 87
Veratti, Emilio 19
vesicles 47–8, 50, 121, 155, 186
villi 48, 133
Virchow, Robert 17
viruses 144, 145
vitamin C 121
Volvox 126, 132
von Gerlach, Joseph 18
von Sachs, Julius 19

W
water cycle 100
water mold 116
water supplies 150, 152
Watson, James 23
Wilkins, Maurice 23

Y
yeasts 71, 86, 116, 146

Z
zygotes 70, 77, 114, 117, 134

Acknowledgments

PICTURE CREDITS

The publisher would like to thank the following individuals and organizations for their kind permission to reproduce the images in this book. Every effort has been made to acknowledge the pictures; however, we apologize if there are any unintentional omissions.

A. El Albani 128; Maggie Bartlett, NIH/NHGRI 24 (center); Biodiversity Heritage Library 20 (left); The Company of Biologists, republished with permission from the *Journal of Cell Science*, 39, 1979, pp. 257–272, figs. 11–14; permission conveyed through Copyright Clearance Center, Inc. 37; Corbis © Pallava Bagla/Corbis 85; Dartmouth Electron Microscope Facility, Dartmouth College 51 (right); Darwin Trust 81 (top); Dr. Phil Dash 163, 176; Courtesy of M. L. Escande and H. Moreau 109 (left); Stephanie Evans 118; FAO 84; Dr. David Phillips/Getty Images 51 (left); Andrew Gooday, National Oceanography Centre, Southampton 109 (right); Illustration by David S. Goodsell, the Scripps Research Institute 106; King's College London 23 (right); Inside a Neuron by Patrick Nahirney and James Tyrwhitt-Drake 28; Nanoscience Instruments 153 (main); © The Trustees of the Natural History Museum, London 80; Polygonal oscillation cracks in the 3.48 Ga Dresser Formation, Pilbara, Western Australia. Photo by Nora Noffke (after: Noffke et al. (2013)) 81 (bottom); George Palade/The Cell: An Image Library–CCDB (http://www.cellimagelibrary.org/images/7605) 27 (top)

Public Health Image Library (PHIL):
27 (center left) Larry Stauffer, Oregon State Public Health Laboratory; 27 (center right) National Institute of Allergy and Infectious Diseases (NIAID); 33 (left), 78 Janice Haney Carr; 98, 99 CDC/James Archer; U.S. Centers for Disease Control and Prevention; 104 (inset) CDC/Dr. E. Arum; Dr. N. Jacobs; (105 bottom) National Institute of Allergy and Infectious Diseases (NIAID); 114 CDC/ James Gathany; 115 (top) CDC/ Dr. Mae Melvin; 115 (center); CDC/ Dr. Stan Erlandsen; 129 CDC/ Dr. Lucille K. Georg; 142 National Institutes of Health (NIH); 145 (top and center) National Institute of Allergy and Infectious Diseases (NIAID); 145 (bottom) CDC/Fred Murphy, Sylvia Whitfield; 146 (bottom) CDC/Dr. Leanor Haley; 151 (main) CDC/Peggy S. Hayes, photographer Elizabeth H. White; 151 (top right) CDC/Dr. Balasubr Swaminathan, Peggy S. Hayes, 151 (bottom right) CDC/Bette Jensen, photographer Janice Haney Carr; 154 (left and right) National Institute of Allergy and Infectious Diseases (NIAID)/David Dorward; 155, 157 National Institute of Allergy and Infectious Diseases (NIAID); 166 CDC/Janice Haney Carr

Dr. Yale Rosen 161

Science Photo Library:
5 (top) Wim Van Egmond/Visuals Unlimited, Inc./Science Photo Library; 6 CC Studio/Science Photo Library; 14 (left) Biophoto Associates/Science Photo Library; 19 (left) Science Photo Library; 19 (right) Biophoto Associates/Science Photo Library; 23 (left and center) Science Photo Library 23; (bottom) A. Barrington Brown/ Science Photo Library; 24 (right) James King-Holmes/Science Photo Library; 27 (bottom right) Wim Van Egmond, Visuals Unlimited Inc./Science Photo Library; 27 (bottom left) Dr. Lothar Schermelleh/Science Photo Library; 30 (bottom left) A.B. Dowsett/Science Photo Library; 30 (bottom right) Eye of Science/Science Photo Library; 33 (right) Sinclair Stammers/Science Photo Library; 32 Dr. Jack Bostrack, Visuals Unlimited/Science Photo Library; 35 Don W. Fawcett/Science Photo Library; 41 Dr. Elena Kiseleva/Science Photo Library; 43 Biophoto Associates/Science Photo Library; 46 Don Fawcett/Science Photo Library; 47 J.C. Revy, ISM/Science Photo Library; 49 (left) Dr. Torsten Wittmann/Science Photo Library; 49 (right) Science Photo Library; 53 Don Fawcett/Science Photo Library; 55 (top) Adam Hart-Davis/Science Photo Library; 55 (bottom) Alfred Pasieka/Science Photo Library; 57 Dr. Kari Lounatmaa/Science Photo Library; 58 Dr. Linda Stannard, UCT/Science Photo Library; 61 A.B. Dowsett/Science Photo Library; 65 (bottom) CNRI/Science Photo Library; 68, 69 Jennifer Waters/Science Photo Library; 71 David Scharf/Science Photo Library; 72, 73 Biophoto Associates/Science Photo Library; 74 AJ Photo/Science Photo Library; 75 (top) Steve Gschmeissner/Science Photo Library; 77 (right) Biophoto Associates/Science Photo Library; 86 Pascal Goetgheluck/Science Photo Library; 88 (top) James King-Holmes/Science Photo Library; 89 Thomas Deerinck, NCMIR/Science Photo Library; 90 Gerd Guenther/Science Photo Library; 92 P. Motta, Dept. of Anatomy, University "La Sapienza," Rome/Science Photo Library; 93 (top) East Malling Research Station/Science Photo Library; 93 (center) Eye of Science/Science Photo Library; 93 (bottom) Biophoto Associates/Science Photo Library; 94, 96 (left and top right), Eye of Science/Science Photo Library; 96 (center) Aurélien Celette, Mona Lisa Production/ Science Photo Library; 97 (top and bottom) Eye of Science/Science Photo Library; 101 Power and Syred/Science Photo Library; 102 (left) Wally Eberhart, Visuals Unlimited/Science Photo Library; 102 (right) Dante Fenolio/Science Photo Library; 103 (top) AMI Images/Science Photo Library; 103 (bottom) NIBSC/Science Photo Library; 104 A. Dowsett, Health Protection Agency/Science Photo Library; 105 (top) Eye of Science/Science Photo Library; 111 (bottom left) Frank Fox/Science Photo Library; 111 (bottom right) Volker Steger/Christian Bardele/Science Photo Library; 111 (main) Wim Van Egmond, Visuals Unlimited Inc./Science Photo Library; 112 (top) Biophoto Associates/Science Photo Library; 112 (bottom) Steve Gschmeissner/Science Photo Library; 115 (bottom) Nigel Cattlin/Holt Studios International/Science Photo Library; 116 (left) Dr. Ken Wagner, Visuals Unlimited/Science Photo Library; 121 Thomas Deerinck, NCMIR/Science Photo Library; 122 Science Photo Library; 124 David Scharf/Science Photo Library; 126 (top) Dr. Peter Siver, Visuals Unlimited/Science Photo Library; 127 (bottom) Ralph Robinson, Visuals Unlimited/Science Photo Library; 132 (top) Dr. Robert Calentine, Visuals Unlimited/Science Photo Library; 135 Professor Miodrag Stojkovic/Science Photo Library; 140 Eye of Science/Science Photo Library; 141 Dorit Hockman/Science Photo Library; 149 Eye of Science/Science Photo Library; 158, 159 Steve Gschmeissner/Science Photo Library; 164 Simon Fraser/Science Photo Library; 167 Steve Gschmeissner/Science Photo Library; 168 Science Photo Library; 170 Herve Conge, ISM/Science Photo Library; 171 (right) L'Oreal, Eurelios/Science Photo Library; 172, 173, 174 Steve Gschmeissner/Science Photo Library; 179 Eye of Science/Science Photo Library; 180, 183, 184 Steve Gschmeissner/Science Photo Library; 187 Eye of Science/Science Photo Library

Shutterstock Inc.:
3 Marioner; 5 (bottom); Jubal Harshaw; 7 (left) Zastolskiy Victor; 7 (center) Anita Patterson Peppers; 7 (right) Lilyana Vynogradova; 25 solarseven; 30 (top right) Kichigin; 30 (top center) Chris Hill; 30 (top left) igotabeme; 60 Denis and Yulia Pogostins; 62 Claudio Divizia; 65 (top); 75 (bottom) Oleg Kozlov; 76 Paul K; 77 (left) Jubal Harshaw; 83 Norma Cornes; 86 Kanjanee Chaisin; 95 Shane Myers Photography; 100 (left) Dmitry Vinogradov; 100 (center) Tawann P. Simmons; 100 (right) Pan Xunbin; 113; 117; 120 (top left, top center; 120 (top right) photowind; 120 (bottom left, bottom center) Jubal Harshaw; 120 (bottom right) D. Kucharski K. Kucharska; 125 (top) Lighthunter; 126 (bottom) Lebendkulturen.de; 130 Merlin74; 144 Nicolesa; 147 Joerg Beuge; 152 (bottom) John Wollwerth; 152 (top) David W. Hughes; 153 (left) Dmitry Naumov, 162 Loren Rodgers; 171 (left) Australis Photograph; 175 (top) Jose Luis Calvo; 175 (bottom) Pan Xunbin; 177 Jose Luis Calvoy

J. D. Taylor, The Natural History Museum 102 (center); Greg Wanger and Gordon Southam 146 (bottom)

Wellcome Images:
18 Bernard Maro/Wellcome Images; 48 (left) Kate Whitley/Wellcome Images; 48 (right) Paul Appleton/Wellcome Images

Wellcome Library, London:
1, 2, 10, 12, 13, 14 (center) 14 (right), 16, 17, 22, 24 (left), 108, 138, 139

Wikipedia:
20 (right), 88 (bottom) © CC-BY-SA-2.0 Toni Barros; 113 (left) © CC-BY-SA 2.5 Richard Bartz; 116 (right); 127 and 131 © CC-AS-A-3.0 Christian Fischer; 131 (inset) © CC-AS-A-2.0 Plantsurfer; 134 (inset) © CC-BY-SA-3.0 Josef Reischig; 135 (inset) © CC-A-2.5 Nissim Benvenist; 169 © CC-AS-A-4.0 Dr. Graham Beards

Chelsea Wood 132 (left)

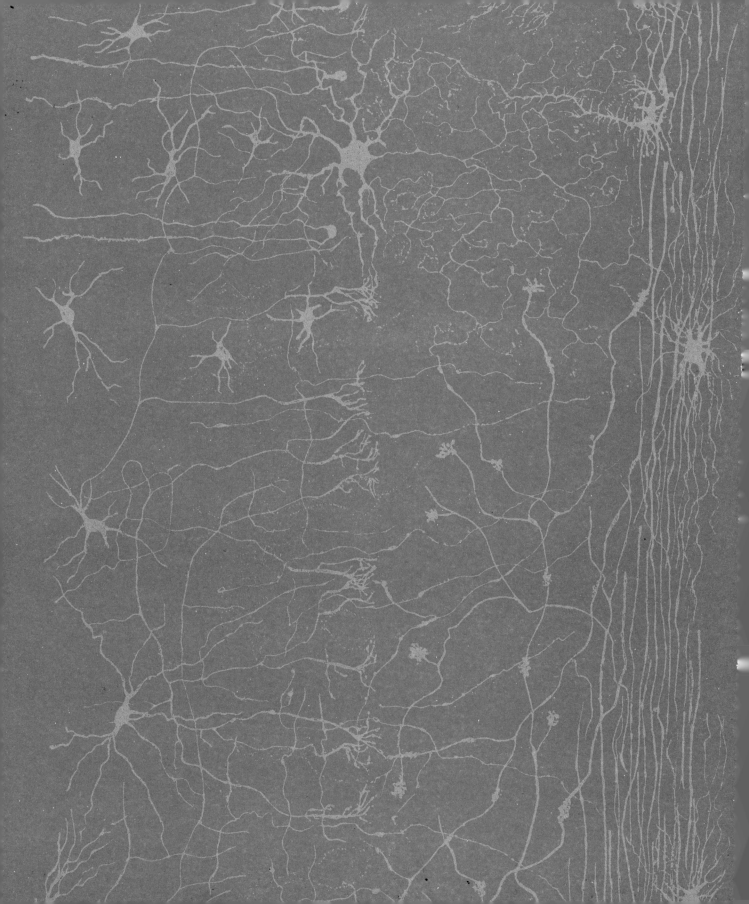